MOBINOMICS

MOBINOMICS

Mxit
and Africa's
mobile
revolution

Alan Knott-Craig
with
Gus Silber

ISBN: 978-1-920434-36-6

First edition, first impression 2012

Published jointly by
Bookstorm (Pty) Limited and Pan Macmillan South Africa
Suite 10   Private Bag X19
Private Bag X12   Northlands 2116
Cresta 2118   Johannesburg
Johannesburg   South Africa
South Africa   www.panmacmillan.co.za
www.bookstorm.co.za

Distributed by Pan Macmillan
Via Booksite Afrika

Edited by Mark Ronan
Proofread by John Henderson
Design and typesetting by Triple M Design, Johannesburg
Printed and bound by Ultra Litho (Pty) Limited

CONTENTS

ACKNOWLEDGEMENTS

W E STAND on the shoulders of those that have come before us. And, of course, those that stand next to us. A sincere thank you to:

All the people that built Mxit. Many years, hearts and bodies were burnt giving birth to this dragon egg, those contributions will always be part of the story.

Gavin Varejes, my first angel. Brett and Mark Levy, the guys that gave me my second big break. Pierre Fourie, my first business partner and my gateway to the best engineering minds in the world. Louise and Terry, for their endless patience. Gus, the translator of the magic. My staff. My mates. My bros. My parents. My daughters. My wife.

Tx, Alan

I HAVE HAD a keen interest in the story of Mxit ever since my teenage daughter came home from school one day, raving about this new thing – that's what she called it, 'this new thing' – that allowed her to stay in constant touch with her friends and schoolmates on her cellphone. As her prepaid airtime sponsor, I was aghast, until she told me what the messages cost: practically nothing. I still don't quite believe that part, but I'm willing to give her and the other multiple millions of Mxit users the benefit of the doubt.

So, some thanks.

Firstly, to Alan, for his tireless enthusiasm, support and unflappable belief in the transformative power of mobile technology. To Alan Knott-Craig Snr, thanks for valuable insights into the world of mobile.

Then, to Louise, Terry, and everyone else at Bookstorm and Pan Macmillan, thank you. To all who gave freely of their time and expertise at Mxit and World of Avatar, my sincere gratitude. In particular, to Gavin Marshall, Marnus and Mariana Freeman, Traveller, Jaco Loubser, Juan du Toit, Michael Carter, Paul Stemmet, Pieter Rautenbach, Craig Whittaker, Johan Jacobs, Philip Wilton and his team, and Marlon Parker and his team at RLabs. Thanks to David Weber and Glenn Alexander.

Also (I'm almost done), thanks to Sphiwe, Maru, Dave, Nikita, Candice, Sarah and Mary for their help and patience. Thanks to Herman, who started 'this new thing'. And finally, as always, to Amanda, Sarah-Jane, Max and Rachel for putting up with me during the writing of this book.

Gus Silber

In these criss-crossing threads are
woven the fabric of a community,
a society, an economy, a nation.
And beyond that, the world itself.
But the technology isn't the dream.
The dream is what you can do with it.

One's company, two's a crowd, three's a social network

In concentric circles of family and community, society and nation, a social network takes shape in a little town in the vineyards of the Western Cape, South Africa.

TUESDAY EVENING in Stellenbosch. It's raining. A soft mist sweeping in from the roof of the mountains, falling like a benediction on the rows of green that will one day be turned into wine. One day.

For now, the traffic in Dorp Street heading out of town is heavy and sluggish, as it always is at this hour.

People who've lived in Stellies all their lives will tell you that the town isn't what it used to be. They'll say it's become a place of bustle and noise, and outsiders rushing in to make a quick buck. I don't know about that. The bucks aren't quick in Stellenbosch.

But one thing I can tell you is that the grid-lock isn't as bad as it is in Joburg, either. For starters, there's the view. The sunlight filtering through the leaves of the old oak trees, casting the curls of the Cape Dutch gables into stark relief. The water swirling down the stone canals that shoulder the streets. Technically, I suppose, they're gutters. But I think you're allowed to indulge a sense of romance in Stellenbosch.

My heart is in this town, and not just because it's where my wife, Sibella, grew up, and where we're bringing up our family too. My heart is here because I'm building something, investing in an ideal, putting down roots. Small acorns grow into sheltering oaks, seedlings become the vineyard. But the harvest of my dreams is something you can't touch or taste, see or feel. And yet it is as pervasive and as vital as oxygen.

The miracle, for me, never fades. I tap on the face of a small device, fashioned from glass and silicon and plastic. The signals are interpreted, processed, and conveyed on radio waves that swarm and cluster in cells, like the chambers of a hive. Someone talks to me, or reads my words on their screen. We connect.

In these criss-crossing threads are woven the fabric of a community, a society, an economy, a nation. And beyond that, the world itself. But the technology isn't the dream. The dream is what you can do with it.

> Small acorns grow into sheltering oaks, seedlings become the vineyard. But the harvest of my dreams is something you can't touch or taste, see or feel. And yet it is as pervasive and as vital as oxygen.

Run a small business. Find a job. Educate a child. Pay your bills. Bank. Run a big business. Learn, teach, share, counsel, build. Connect.

In Africa, my home continent, there are about a billion people, and more than half of those hold in their possession the most powerful instrument of social, economic and political transformation ever invented. We skipped the printing press, the telegraph and the fixed line, leapfrogging every major breakthrough in communication technology, waiting for the ultimate innovation.

The mobile phone.

Millions of nodes on a network, each representing a person with a dream of their own, and the means, however remote, to achieve it. This is our age of digital Uhuru, of the African mobile economy on the move, and it excites and inspires me to be a part of it.

But this story of mine begins with a networking exercise of the old-fashioned variety. An amble along the banks of the Eerste River in Stellenbosch one August morning in 2011. It was raining. I was sharing an umbrella with a guy named Herman Heunis.

Earlier, I had popped him an SMS: 'Keen for a 9.30 am coffee-walk?' Ah, the coffee-walk. A fine Stellenbosch tradition. There is something about walking in a town like this, something about a cup of coffee to go, that stirs the air and quickens the senses.

You're thinking on your feet, going places, open to serendipity and the elements. You're on equal footing. Which is important, when the overriding item on the agenda for your coffee-walk is a business decision that will dramatically change both your lives. Herman and I had an understanding.

A memorandum of understanding, to be precise, under the terms of which, a few months earlier, I had made an offer to buy a company from him and his equity partners for just under R700 million.

A company called Mxit, an instant-messaging and chat service that in less than a decade had grown to be the biggest social network in Africa. More than that, it had become a cultural force, a community of millions, with its own economy, its own infrastructure, its own systems, its own traditions.

I had become an evangelist, an unpaid, unofficial

ambassador for the network, flipping the lid of my laptop to show anyone who had a moment that the numbers were mind-boggling and that this was so much more than a forum for teenagers to chat and flirt.

It was a marketplace of products and ideas, a platform for lobbying and mobilising, a springboard for education, opportunity, social upliftment, empowerment.

I had begun to think of Mxit as a country, rather than a network – a country with connections all over the world, and its heart beating strongest in Africa. Of course, I wasn't just flying the flag for goodwill and PR. I wanted to raise awareness, and turn that awareness into bundles of cash.

I wanted to see the Mxit flag flying over my fledgling group of mobile-internet startups, World of Avatar. I wanted to live in the Republic of Mxit. I wanted to raise the standard high. We walked. I took a sip of coffee. Bitter.

We talked for a while about a subject that was marginally off the agenda. Religion. I told Herman I had stepped impulsively into a chapel on the way, to say a little prayer. Some deals, you need all the help you can get. He laughed.

Herman is a wiry, no-nonsense kind of guy, with a scraggle of silvery, shoulder-length hair and an edge of flint to his voice. He grew up on a farm in Namibia, a son of the soil who is at his most animated when sharing his passion for the African landscape. He mountain-bikes. He kite-surfs. He 4 x 4s.

A few years ago, roaring through the Namib Desert on a motorbike, he almost wiped himself out in an accident that left him with a broken back and a spell of several months in rehabilitation.

But there is an aura of Zen about Herman too, the soul of an artist, and he grappled over whether to do an arts degree or a

B.Comm. at Stellenbosch University, after his stint of military service, writing software for the South African Navy. He chose commerce.

Computing was his hobby, but it became his obsession, and then it became his fortune. He started out as a tamer of mainframes, the mighty beasts that crunch numbers and mull over data at banks, insurance companies and other mega-corporations.

Then, like an astronomer switching his gaze from Jupiter to a speck of stardust, he scaled down, down, down to the micro-cosmos of mobile. He spun off a division of his mainframe consulting company to build text-based games for cellular handsets.

From that fusion of big and small thinking, plugged into the right technology in the right place at the right time, Mxit was born.

There we were, two guys in brown leather jackets, sharing an umbrella, having a break-up conversation in the rain. Bizarre, I know. But to his eternal credit, Herman was cool about it.

Herman wasn't the coder, the Grand Designer, but he was the founder and the architect-in-chief, and he had assembled a team of wizards and whizz-kids who had come up with something simple and magical, a portal to networks and communities and worlds.

But over the last few years, he had fallen out with his one-third partners, the multinational media conglomerate, Naspers, and he was tired. He wanted to cut his ties and move out of the building. He was a willing seller. I was an eager buyer. But it takes more than good intentions to build a bridge between those stations. We stopped.

'Sorry, dude,' I said. 'I just haven't been able to raise enough cash.'

There we were, two guys in brown leather jackets, sharing an umbrella, having a break-up conversation in the rain. Bizarre, I know. But, to his eternal credit, Herman was cool about it.

We walked back to his car – a brand-new Porsche – and shook hands. No words. He got inside and drove off. I carried on walking, watching the rain chasing the leaves in the gutters of Stellenbosch.

To make matters worse, it wasn't the first time I'd had coffee with Herman and found myself walking away from the opportunity of a lifetime. In 2006, while I was running a wireless broadband service provider called iBurst, I had flown down to Stellies on a mission to stake my claim to Mxit. A modest stake, but still.

The service at that stage had some 360 000 users, more than two-thirds of whom were in the dream marketing demographic of 12 to 25. And more than 10 000 new users were signing up every day.

Even so, it would have been crazy, back then, to project that Mxit would grow to almost 50-million users in 120 countries, 23-billion messages a month, and up to 50 000 daily signups, by 2012. But that's exactly what happened.

Herman sat across the table from me, giving me a look that said, who's this lightie trying to get in on my business? Then he levitated a figure that made me laugh, twice, quietly to myself.

Once back then, because of the sheer audacity of what he was asking, for the size of the stake. Once, now, because in retrospect, what he was asking was a pittance. Thirty-five million rand.

In retrospect you rationalise to dull the pain of regret, but the timing wasn't right for me, and I was still trying to get my own startup off the ground. Coffee, handshake, thank you. I paid the bill and left.

Now here I was again, five years later, walking away, wandering, lost in thought. The stakes this time had been so much higher, the fall so much further. I hadn't felt this let down or defeated since high school. Glen High, in Pretoria East. I must have learnt a few things there – my love of writing and history, and enough basic maths to scrape through university, years later, as a chartered accountant.

But it was a rough-and-tumble institution, with lots of fighting and unruliness, and the biggest lesson I learnt, looking back, was How Not to Get Caught with Alcohol on School Grounds. I learnt that lesson, I'm sorry to say, by getting caught with alcohol on school grounds.

I was in Standard 9, and my misdemeanour earned me a suspension. The only reason I wasn't expelled outright was that my mates didn't rat on me. Three other guys were expelled, each of them ratting on the other. So, don't take booze to class, and don't rat.

My dad did not appreciate the subtle distinction between suspension and expulsion when I told him what had happened, and therein lay another lesson – given by one Alan Knott-Craig to another. He wasn't impressed. And he knew a thing or two about handling wayward children.

As a computer programmer working for the South African Post Office in the 1970s, my dad earned barely enough money to pay the rent. Then an offer came up, and he and my mom,

Janet, a schoolteacher, moved from their tiny flat into a spacious, rent-free home in Pretoria. There was just one catch. It was the Louis Botha foster home in Wonderboom, a place of care for orphans and children from broken homes. My folks took on the job of house-parents, and for the first three years of my life, I grew up with a bunch of kids who had some serious mental baggage.

But the videos show a happy and laughing brood, myself among them, so the house-parents must have done a good job. So much so that when I was ten years old, I went out and got a job of my own. Just part-time, showing people to their seats at my local Nu Metro, and then packing fruit and veggies at the Spar.

> My dad did not appreciate the subtle distinction between suspension and expulsion when I told him what had happened, and therein lay another lesson – given by one Alan Knott-Craig to another. He wasn't impressed. And he knew a thing or two about handling wayward children.

At school I ran the soccer betting pool, and sold icies at break. I wish I could say I invested my micro-earnings wisely, and learnt valuable lessons about entrepreneurship. The truth is, I spent every penny on fantasy books and computer games.

I would sneak into the study after everyone was asleep and play Hero's Quest for hours on end, escaping from dungeons, ascending spiral staircases, accumulating weapons and artifacts, confronting evil wizards, battling fearsome monsters, casting spells, opening secret doors, vaulting over spear-traps, dodging falling rocks, and hoping like hell that my dad wouldn't walk in, tap me on the shoulder, and ask what I thought I was doing on the computer at four in the bloody morning.

'Isn't it obvious? I'm overcoming a series of obstacles while searching for the treasure that will fulfil my quest and make me a hero.'

That would never have worked with my dad. He doesn't dwell in the kingdom of metaphor and analogy. He works in the world of metal and fire, of character forged in the crucible of toil and sweat and grit. And here I'm just talking about what he does for a hobby, making lovingly hand-crafted barbecues. He's an electrical engineer, and that kind of thing comes as easily to him as striking a match. Me, I'm not a natural spot welder.

But it was a ritual of my childhood, my old man calling me over to take the welding gun in my bare hand, and then standing and watching as I touched the white-hot rod to the grid, squinting against the shower of sparks, grimacing at the fiery crackle and the smell of scorched electrodes.

But at the end of it, if all went well, if the spots were neat and the bars of the grid were straight, I would feel like a hero in the eyes of my dad. He's always been a fast talker, a sharp shooter, impatient with needless rules and niceties and excuses for not getting things done.

Someone who worked for him once took a look at his initials on a memo, and turned them into a nickname that stuck: AK-47. That caused great amusement in our household.

In 1990, soon after the release of Nelson Mandela set in motion the waves of change that would shape the new South Africa, my dad was seconded from his management position at Telkom, the state-owned telecommunications company, to help get a new revolution into gear. Cellular radio telephony, it was called – liberation from the tyranny of the landline. The cutting of the umbilical cord that tethered people to one fixed point on a map. We were going mobile.

I was still in high school, trying to figure out what to do with my life, when my dad was appointed chief executive officer of the startup company that would become Vodacom, the national cellular service provider. My grandfather, also Alan Knott-Craig – I'm often referred to as Alan Knott-Craig Jr, but I'm really Alan Knott-Craig 3.0 – ran a small chain of country newspapers in the southern Cape. So I've got ink as well as radio waves flowing in my veins. But when the time came for me to choose a career, I was given the option of pursuing any university course I wanted, as long it was a B.Comm. in accounting. It didn't inspire me, but that wasn't the point. My dad insisted. He said he wouldn't pay for any other course, and then he promptly sold his car to help finance my tuition.

There was a lot riding on my performance, and you could say I passed with flagging colours, just scraping through with a rewrite to get into the honours class. Miraculously, thanks to a seasonal shortage of clerks, I managed to sneak onto the articles list at Deloitte in Cape Town, where I met another chartered accountant who would soon become my wife. Sibella Bosman. Articles was a great time in my life. Living in a digs of five guys, earning actual money, having a party in the Mother City. Work was fun, too. I spent most of my billable hours thinking of ways to escape the office early, and figured out some great new methods of auditing, reducing effort by 50 per cent and producing the same results.

This did not mean that I produced double the output. The

extra time I generated was put to good use on the golf course and on the water, sailing. I wasn't planning on working my way up the ranks to the hallowed status of partner, so shining at work was never a priority.

I was almost fired twice. Once for bunking work, and once for being on a porn email distribution list. At 25, I was done with articles and married to Sibella. We went off to New York, where we worked for a stint before selling up and backpacking around the world for six months.

Back home, reality waited, in the form of the huge Absa loan that had been used to fund our overseas adventures. My assets were souvenirs and memories and the equity of love. But I needed a job. I started phoning around and going for interviews. I didn't want to be a CA. I wanted to work in cellular telecommunication.

Trouble was, I couldn't figure out how to do that without working for one of the networks: Vodacom, which would have meant working for my dad, or MTN, which would have meant … well, I don't know if they were going to be too happy to hire a guy named Alan Knott-Craig.

In many ways I think that I'm a lucky guy. Not that I was born into luck (which I was), but just because luck is a matter of opportunity and circumstance, and sometimes, in my life, they've collided.

One day, out of the blue, I got a call from a man by the name of Gavin Varejes. He asked if we could have a coffee. Funny how no one ever asks, can we talk business? It's always, can we have a coffee? And why not? We had a coffee.

Gavin was an entrepreneur who had purchased the licence for location-based cellphone tracking software from a UK company. Now he was looking for someone to take the licence and turn it into a viable proposition. At our second meeting, he offered me 10 per cent equity and a nominal salary to start Cellfind. Of course, I knew that part of the attraction was my dad's position at Vodacom.

I asked my dad what I should do. He said I should join a big company, rather than try to start a small one. But he wasn't going to stand in my way. I asked Sibella. She also had dreams of entrepreneurship, but was willing to put them on hold for me. She said go for it.

Sibella paid the bills for two years, while I learnt the hard lessons that my dad had warned me about. In fact, after nineteen months I told Gavin to shut the business down. He told me to back myself.

In month 25 we paid our first dividend. In September 2007 the company was valued at a few hundred million rand, and Blue Label Telecoms paid me cash for my stake. That was November 2007, just before the global financial crisis. I got lucky.

I was already running iBurst at that time. It was a licensed broadband operator, and Blue Label Investments had bought 40 per cent of the business blind – no due diligence, nothing – before sending me in to fix things up. The deal for me was no equity, only a salary, and the opportunity to build something big. I signed up like a marine and stormed into battle with my troops.

We took a company with no share register, no customer care, no billing records for 15 000 customers, no audited accounts for five years, 50 employees and a physical network of 40 rag-tag base stations, and somehow turned it around.

After three years, we had 120 000 customers, a national network of 350 base stations, a history of unqualified audits, the MyADSL prize for Broadband Provider of the Year, and the Deloitte Best Company To Work For prize for our industry.

I must admit it went to my head.

Despite the successes, however, iBurst could not escape mediocrity. We talked a good game, but simply couldn't get to the next level. My dad had warned me about the difficulties of starting your own business. What he hadn't warned me about was the cost.

During the iBurst years, I drifted apart from Sibella. I was fighting a roaring battle, fending off competitors, managing shareholders and motivating staff. I would get home late in the evening, only to be told that I wasn't giving my family enough attention. I felt like I was walking a tightrope across a raging chasm.

> Despite the successes, however, iBurst could not escape mediocrity. We talked a good game, but simply couldn't get to the next level. My dad had warned me about the difficulties of starting your own business. What he hadn't warned me about was the cost.

I visited my dad and asked how he managed to deal with balancing the demands of work and family, my implication being that work is so much more important. He had just reached a milestone birthday, and he was in a reflective mood. 'When you're 60,' he said, 'it's not your employees that are around you, it's your family. Look at your family and look at your staff. And choose.'

At that point, he'd had two heart attacks and a divorce, so I guess he was speaking from experience. For once, I could see my way to the other side of the chasm.

I went to see Brett Levy, my chairman at iBurst. I knew he

wasn't going to be happy with what I had to say. But he understood. I resigned so that I could move with my family from Joburg to Stellenbosch, and put some effort into rebuilding my own troubled network for a change. But first, the handover.

Four months of pain and trauma. Many people felt betrayed, angry and hurt. In the process of trying to avoid one divorce, I had plunged straight into another. No matter how much I tried to justify it, it looked like desertion in the heat of battle.

I learnt a bitter lesson: never gather followers unless you're prepared to lead them in the front line for a hundred years. I had made my choice.

We settled in to Stellenbosch, which at least offered a change of pace and place. Then, one Sunday in January 2010, I picked up the *Sunday Times* to find a familiar pair of names all over the front page. Alan Knott-Craig and Alan Knott-Craig.

It was a hard-hitting story, rife with allegations that the CEO of Vodacom had benefited or favoured family members in awarding a contract and assisting with business ventures. One of those family members, so the story said, was me.

The 'Alan Knott-Craig nepotism scandal' dominated the headlines for days, and although the board and shareholders of Vodacom exonerated my dad of any wrongdoing after an independent enquiry, it hit very close to home.

By nature I'm an optimist, but that spell in the spotlight left me feeling embarrassed and defensive, and I decided to lie low for a while. A couple of months later, I boarded a plane to the US, for a ten-day course on public policy and leadership at Harvard, as part of the World Economic Forum's Young Global Leader programme.

Far away from home, far from iBurst, far from the mess of it all, I had an epiphany. There were two things, I realised, that were more important to me than anything else. The first was family, followed by the pursuit of my purpose in life. I just had to figure out what that was.

On the journey home, a series of hops that kept me in the air for 32 hours, a word popped into my head as I gazed at the rolling clouds. 'Avatar'. An earthly manifestation of a divine being, a spiritual guide, a dream-warrior on a mission. And the image you choose to represent you or your alter ego on video games and social networks.

By the time we touched down, on April Fool's Day, 2010, I knew exactly what I wanted to do. I had found my purpose: to help people make money and build community via their mobile phones.

> There were two things, I realised, that were more important to me than anything else. The first was family, followed by the pursuit of my purpose in life. I just had to figure out what that was.

I asked one of my original technical partners at Cellfind, Peter Matthaei, to visit me in Stellenbosch and we spent a weekend walking through the vineyards talking about the world and tech in general. It was a conversation about chasing greatness and doing something meaningful. The seed was planted, and Pete and I were ready to nurture it into to World of Avatar.

Things began to move at the speed of lightning. Pete's development team moved to Stellenbosch, bringing Mobicanvas, a tool for building cheap mobile websites, with them. Steve Briggs called in a panic and I stepped in to save Arc Telecoms, a voice and data service provider.

Tom London, a lanky, laid-back talk radio presenter, moved

from Johannesburg to pursue his plans for an internet-based radio station. Veeren Naidoo quit everything to chase his dream of an online bill-payment service called Triloq.

Johnny Graaff approached me to get involved in FSMS, an ad-supported web-and-mobile SMS service. Katy Digovich flew in from Gaborone and we agreed to launch Jujuz to sell mobile classified ads.

Andrew Rudge and I bonded over many a coffee and we decided to try MobiFin, a platform for distributing financial information and other educational content on mobile devices.

Pete Matthaei, my co-founder and chief technical officer, wasn't enjoying being pulled in ten different directions. He wanted to chase his dreams of building a world-changing product. Which turned out to be BOOM!, an online music discovery service.

A close friend of mine, Nico van der Westhuizen, moved to Stellenbosch in April. He and I had both started our first businesses in 2003, sharing many of the trials and tribulations of first-time entrepreneurs.

Immediately upon settling in, he started persuading Kevin Harris and me to establish a passive investment holding company to help us diversify our assets and to learn new industries. After much umm-ing and ah-ing, we agreed to give it a go. The fateful meeting was at Basic Bistro on Church Street, so we decided to call the company Basic Business Investments (BBI).

Through one of Sibella's best friends I met Jimmy Hanekom, a typical disgruntled employee who felt he could do a better job than his boss, if only he had his own company. I backed him. Problem was that it was a property-development company, and you can do squat without a balance sheet.

Nevertheless, he cracked on, and that company became

Foundation Capital. My shares were bought by a good friend, Johan Bosman, and BBI.

In the online media arena, I invested in a venture run by one of the most driven and determined mavericks I know. Branko Brkic, editor and publisher of the *Daily Maverick*, a journal of news and commentary that is helping to shape the way we see our noisy, disruptive democracy.

Before I knew it, we had a group of rockstars – or avatars, as I would come to call them – and no way of going back.

Of course, we had a couple of hiccups. I was introduced to a woman with big dreams of rolling out IPTV (internet protocol TV) and completely disrupting the pay-TV industry. Music to my ears. After giving her R500 000 in good faith, things rapidly started going pear-shaped. I decided to walk away when it was clear she couldn't work with my guys, and worse, was insulting them.

I also made the rash decision to start an office-letting-cum-tech-incubator and signed a five-year lease for 500 square metres of the most expensive office space in Stellenbosch.

Meanwhile, on the home front, Sibella and I began working together to make our marriage work. To my everlasting surprise and relief, we identified the problems, dealt with them, and realised we were still crazily in love.

Thank God, because that's when the real business pressure began. Without a rock-solid relationship at home, I would probably have crashed.

It all happened in August 2010. On the first of the month, I received an email from my banker, saying the bank had changed its mind regarding the pre-approved additional facility on my house, thereby removing R5 million of capital from my war chest. At that point, I had R1 600 remaining in my overdraft, and an overheads bill for World of Avatar alone of R250 000, due on the 25th.

On 12 August, I re-proposed to Sibella. She accepted, luckily for me.

On 13 August, a Friday, we opened the first World of Avatar office and had an official launch party.

On 14 August – my birthday – Sibella informed me she was pregnant again.

On 15 August, I double-checked my bank balance. Yup, still R1 600 left.

On 16 August, I finally plucked up the courage to email my dad and ask for an emergency loan. He mulled over it for a couple of days, and then said he would help me with 20 per cent of the amount. And I had better pay it back in a hurry.

I borrowed money from friends. I found money I'd forgotten about in an overseas account. I searched like a chicken pecking in the dust. And then one day, I went to a kiddie's birthday party in Stellenbosch, and bumped into an angel. The best kind of angel: the kind that, with a little bit of persuasion, puts faith, trust and money into your business dreams and ventures. To my eternal disbelief and relief, I'd found a capital partner. On a handshake, I was saved.

I got lucky.

One day in early January 2011, in Stellies, I was having a casual breakfast with one of my closest friends, Francois Swart. Francois, aka Frank, was CEO of a listed private equity company called Paladin, based in Stellenbosch.

Just like that, Frank asked whether I would consider bringing him on board the World of Avatar train. I was flabbergasted. Frank was always the top-rated clerk at articles, seventh in the board exams, and a rockstar at Goldman Sachs in London. And here he was saying he was keen on the World of Avatar story.

Clearly, I can talk a good game. Over the next few weeks, we thrashed out a deal whereby he bought 20 per cent of my shares and could wriggle his way out of the PSG Group without offending anyone.

This proved to be quite serendipitous because my personal finances were not looking hot. I was a classic case of 'asset rich, cash poor'.

We sealed the deal in February 2011, and by the end of March, Frank was in the fire. I'll never forget when he asked whether I minded if he wore baggies to work. This was his first experience of working outside a corporate. I said, please don't ask anything like that again. He came to work in baggies. Suffice to say, I love Frank.

The businesses were ticking over, with a couple of rockets taking off, but I still felt there was something missing from the mix. I popped into an office just outside town. The office of Herman Heunis, founder and CEO of Mxit.

Hello, Herman. Me again. This was in April 2011, four months before we would take that walk along the Eerste River. After an hour of pleasantries, I casually suggested to Herman that I'd like to throw my hat into the ring, should Mxit be for sale. He laughed. Turns out I was serious.

He called me a couple of weeks later, and after a few more weeks of negotiation, we signed a formal memorandum of understanding on 11 May, for World of Avatar to acquire between 90 and 100 per cent of all the Mxit companies for R669 million. How did I get to the valuation?

Well, when I started the negotiation with Herman, I had to put a number on the table. I knew that number had to be sufficiently big to get him excited, otherwise (a) he would not take me seriously, and (b) he would look around for other buyers to up the price.

My gut said anything less than a billion was a steal. Herman owned 60 per cent of the company. So I came up with a valuation of R669 million, which meant his share would be a nice round R400 million, which sounds better than R390 million. Retail Psychology 101.

Then started the most exciting and stressful four months of my life. We pulled together a star team, conducted a full due diligence within four weeks, held two strategy sessions with all the big hitters present, appointed RMB as our corporate advisors and went on a two-week roadshow to investors.

And failed.

Everyone loved the story. No one loved the numbers. We eventually agreed on an extension. The big day arrived. I didn't have the money. It probably didn't help that Greece was going bankrupt, the US government debt was downgraded and global stock markets had their biggest fall since 2008.

That aside, our failure to raise the cash was an indictment of me. Potential investors didn't believe I could make the magic happen.

I felt like a loser.

I woke up on 19 August, with a sick feeling in my stomach. I sent Herman an SMS. 'Keen for a coffee-walk at 9.30 am?'

Well, some walks take you further than others.

I felt crushed, but I psyched myself into believing there would be other possibilities, other opportunities. I'd lost Mxit, but life would go on. I woke up that Monday feeling on top of the world for some inexplicable reason. Then my attorney, Jan Viviers, phoned. 'Alan,' he said, 'the deal might be alive.'

In a nutshell, his advice was to make a cash offer and

see what happens. Since May, Naspers and Herman had lost patience and energy. They had emotionally disengaged, and World of Avatar was far and away their preferred and most likely buyer. So maybe, just maybe, they would accept a lower price. But where to find the cash?

Long story short, I found enough to make an offer to purchase Mxit for R330 million. A discount of more than 50 per cent on the original offer, but better than nothing!

At 4 pm on 1 September, I sign the letter in my attorney's office. Ten days for the sellers to accept. Jan and I finish a quick beer, and I head home for some play time with my girls.

Sibella and I dash off at 5.30 pm for the launch of Livewell Suites, my buddy Jimmy Hanekom's new venture. It's a series of lifestyle and healthcare facilities for senior citizens. Very smart and stylish. I tell Jimmy that if things don't go according to plan with World of Avatar, I might have to move in.

Friday. A long day. No word from the sellers.

Weekend is *lekker*, just chilling with the girls.

Monday, no word from the sellers. I'm getting nervous.

Tuesday, I get an email from the sellers' attorney, requesting an extension in order to consider the offer more carefully. Golf in the afternoon with Jan. He shakes his head. We can't give an extension.

September 10 comes and goes. Now what? I tell the sellers we're getting deal fatigue. The ball is in their court.

Four days later, they call to say a signed agreement is on its way. We have 24 hours to sign, or they go overseas to find a buyer.

At 4.55 pm on Friday 16 September, we put pen to paper. Price: officially non-disclosable. But a whole lot less than the original R669 million. This deal was meant to be.

And now here I am again, on a Tuesday evening in

Stellenbosch, walking in the rain, crossing Dorp Street, smiling to myself, doubling my pace across the courtyard, and slipping into the sanctuary of Gino's, the biggest pizza-and-beer hangout for students in this student town.

Except, there are no students here tonight. Just us. We've booked the whole place out. We're celebrating. Mavericks, coders, dreamers, wizards, whizz-kids, techies, accountants, avatars and me – an upstart restarting a startup.

We're sitting at the long wooden banquet tables, talking, laughing, shouting over the hubbub of clinking glasses and the waft of hot dough rising in the pizza oven.

And somewhere above it all, silently, invisibly, the signals are shooting like stars, from handset to base station, across thousands of kilometres of cable, under the sea, onto a server on a distant shore, and all the way back again, over and over. People sending messages to each other. People communicating. People searching, reaching out, networking, connecting.

Welcome to Stellenbosch. Welcome to my world. Welcome to the New Republic of Mxit.

IF YOU SEE A FORK IN THE ROAD, BEND IT

CHAPTER TWO

Building a mobile community calls for persistence, innovation, and a natural ability to conjure up an alternate reality. The story of how Mxit came to be.

G AVIN MARSHALL stood in his office, a coder's den of books and disks and coffee cups and scatter cushions and scribblings of Big Ideas and Things To Do Before Lunch. He was looking for something sharp, something to drive home a point he'd been making about the connection between computer programming and magic.

He furrowed his brow, as if trying to reboot his memory, and for a moment he looked frozen, like a character in a computer game, awaiting further input. Then he retreated to the kitchen, returning a few seconds later with the object of his quest held aloft. A fork.

He handed it over for the ritual sceptic's inspection. It was just an ordinary table fork, generic in design, not too hefty, but not too pliable either. Gavin took the fork and pinched it between thumb and forefinger, shaking it, relaxing it, until it looked as blurry as a propeller in the air.

He held it upright and began rubbing the back of it, slowly, all the while talking about how fragile our perceptions are, and how the real magic of things lies in the way our minds trick us into believing that things are magic. The fork melted in his hands.

He twisted it like a piece of licorice, and curled the middle tines forward, into a rock 'n' roll salute. Then he handed it over as a souvenir. The fork had not just been bent – it had been sculpted, coiled into the shape of a cobra with head raised and fangs ready to strike.

So how could this piece of solid metal, its molecules shifted and reconstituted with no apparent effort, be considered an illusion? The conjurer smiled. That was exactly the point.

We see what we want to see, building our own mental images of the worlds around and beyond us. We communicate on networks constructed from strings of code, connecting with each other on invisible waves of radiation that travel at the speed of light.

We hold pictures in our heads of friends we have never met. We gather disconnected scraps of information, and weave them into new stories and mythologies. We tap symbols on keypads to express our emotions. Reality, says Gavin, is a lot more malleable than we think.

Gavin is head of innovation at Mxit. He has been here since its genesis, a benign, easy-going figure with an understated air of authority and a natural flair for holding people in his sway. It is a skill he picked up, along with the odd magic trick, while studying for his Licentiate of Theology.

He practised as an evangelical minister for five years

– Pastor Gavin – and also, occasionally, as a clown, spreading the word in public with the Scripture Union.

What he liked about that was the paradox of engaging and communicating openly with an audience, while keeping his true identity at bay. Behind the face paint and orange wig, he could be anonymous, exploring his real self by pretending to be somebody else.

The truth is, we all have multiple selves, multiple 'profiles', switching subconsciously between them at work, at home, among friends or strangers, according to the circumstance or social network. Gavin reflects on the meaning of this for a moment. 'Actually,' he says, 'I was a clown when I first met my wife.'

'I've always believed that if you put enough bugs in your code, you can generate enough chaos to create life,' he says. Then he laughs. It's an old coder's joke.

Later, he grew disillusioned with the narrow boundaries of monotheistic doctrine, and he left the church to look for truth and faith in faraway places. He took a sabbatical to Peru, where a shaman, a traditional teacher and spiritual guide, opened his eyes to new schools of myth and magic and belief.

Back home, Gavin became a kind of shaman himself, a weaver of spells, a facilitator of altered states of awareness. A computer programmer. He taught himself how to code, learning the languages and protocols that can stir inanimate machines into conscious states of grace.

The mathematics of it came easily to him, because he is a musician too, a guitarist and jazz drummer, keeping time to strict notation and breaking free to improvise, to embellish, to turn the melody inside out and lead it to unexpected places. He mulls over the suggestion that the rigorous syntax of computer code must surely leave little room to sketch outside the lines. Not necessarily so. He thinks of coding as a creative act, like sculpting or music.

'I've always believed that if you put enough bugs in your code, you can generate enough chaos to create life,' he says. Then he laughs. It's an old coder's joke.

In 2004, ten years after the arrival of the cellular phone in South Africa, Gavin was part of a small team of developers working at Clockspeed Mobile, a research and development company based at Riesling House in Stellenbosch. Their assignment was to invent games and other applications that could take advantage of a new revolution in the way people were using their mobiles.

Beyond the everyday miracle of the voice-to-voice call, and the convenience of being able to connect with someone as they moved from location to location, you were now able to send a short message – an SMS or 'text' – that would land in your inbox with a beep.

If you had the right phone, and the patience to learn how to use it, you could browse news headlines and access your email, using a bare-bones protocol called WAP (Wireless Application Protocol, or Wireless And Pointless, as it was known to its detractors). And, crucially, you could play computer games, right there on the screen of your phone.

The CEO and sole funder of Clockspeed was Herman Heunis, a software engineer of the old school, back in the days when punch cards were fed like sacrificial offerings to giant, brooding mainframes. Now his chief enterprise, at the helm of Swist Group Technologies, was the management of massive databases and the provision of support services to cellular telecommunications companies. But he had smaller ambitions.

Herman was in the air one day, whiling away the time by playing a back-of-the-seat game on an in-flight entertainment console, when he had a sudden Archimedean vision: wouldn't it be cool if every passenger on the plane could simultaneously play the same game in a knockout tournament that would end with only one winner?

Then he applied that line of thinking to a community of wireless devices, hundreds of thousands of people connected by a single-minded pursuit, across a mobile network. Eureka. The game was born, and it was called Alaya.

You were the pilot of a rocketship, heroically steering your way through the cosmos, dodging asteroids and space debris, in your quest to be first to touch down on the mysterious planet of … Alaya. The truth of it was a little more parochial. You were the user of a mobile phone, painstakingly tapping out instructions to navigate your ship by SMS: Go straight, turn right in 15 metres, go faster, turn left … beep, beep, beep.

There were no graphics, and you had to plot your calculations and trajectories on paper before making a move. The game would last a full week, from Sunday to Saturday evening, and the SMSs would be sponsored by a big global brand like Ferrari. A million players, all playing the same game at the same time on their mobile phones. That was what Herman had in mind.

But he couldn't convince a sponsor, global or local, to add a degree of money and allure to a game that felt more like a science exam. Alaya blew up in space, within a few weeks of launching. Total number of competing multiplayers: about 40, some 999 960 short of the target. But the game wasn't over.

The big difference between a true entrepreneur and a non-entrepreneur, Herman once told an American business forum, is an almost pathological level of persistence, an 'illness' that keeps you going in the face of facts and logic. You're burning money, your projects are crashing, but you put off pulling the plug until tomorrow, because, as the Afrikaans saying puts it, *môre is nog 'n dag.* Tomorrow is another day.

You can't shut down on a Thursday, because tomorrow is the weekend. And when Monday

'Entrepreneurs have unreasonable expectations about the next big thing. They will ignore advice and reality. They will just believe that somehow, they are going to change the world.'

rolls around, you think, let's give it a few more days, and see if anything comes up. You can't shut down on a Wednesday, because tomorrow is Thursday, and … you keep postponing the inevitability of surrender, in the dogged belief that you need to leave a little gap for the inevitability of success.

Wynand Coetzer, a business consultant and professor of electrical engineering, who first met Herman at Stellenbosch University, calls this the syndrome of the unreasonable entrepreneur. 'Entrepreneurs have unreasonable expectations about the next big thing,' he says. 'They will ignore advice and reality. They will just believe that somehow, they are going to change the world.' He wouldn't have thought, back then, when Herman was working as a systems analyst at the computer centre on campus, that this particular postgrad had the makings of a great tech entrepreneur. But who knows? 'Progress,' he says, 'depends on unreasonable people.'

Back at Riesling House, after the crash of Alaya, Herman's team of software devs were undaunted. The code had been written and the premise put in place for a mobile application that used messaging to connect and build a community. It would take a lot of tweaking and splicing and re-engineering, but in the meantime, there were games to be played.

For Jaco Loubser, fresh out of his computer science degree at Stellenbosch University, this was a dream assignment. Design and develop the mobile games that you yourself would want to play. He worked on a couple of racing-car games, using simple sprite-based graphics of the variety found in Pong and Space Invaders.

> You would wear the tag in public, and if there was a match within range, your DateMate would buzz and vibrate.

Then came Leonardo da Monkey, a two-dimensional scrolling shooter that was slick enough to be sold to China Mobile, the biggest mobile telecommunications company in the world.

'One of my goals when I was at university was to be part of something that could influence people's lives,' says Jaco. 'I wanted to develop something that would actually matter.'

This wasn't it ... yet. Gavin, meanwhile, was making the switch from coding and scripting for the World Wide Web, to programming for the mobile phone, a challenge akin to rigging the sails of a small ship through the neck of a bottle. The size of your code is dictated by the size of the device, and you have to break a few rules and abandon a few conventions to squeeze it all in.

Gavin was happy with this. Constraint inspires innovation. Think of a haiku, which must consist of three lines of five, seven and five syllables, and must contain the germ of an idea that can enlighten the universe. Small is beautiful.

Gavin wrote a game called Kung Fu Master, combining his passion for the martial arts – on his Peruvian retreat he had practised kung fu with the shaman as part of his routine – with the reach and power of a device that puts the world in the palm of your hand. The benchmark phone for the mobile gaming developer, back then, was the Nokia 3510i, a slender slab with chiselled lines, which gave it a slightly military look.

But it had a high-resolution colour screen, internet connectivity, picture messaging, SMS chat, Java games and applications, and two-tone replaceable covers. It was a sexy phone. Even computer programmers, though, must concede that there is more to life than programming sexy phones.

So the Clockspeed team designed a device called the DateMate, a little tag with a radio frequency, which you programmed with your personal details and the profile of your perfect partner. You would wear the tag in public, and if there was a match within range, your DateMate would buzz and vibrate.

It was an early variant of the principle of social networking, or, as it was then known, 'meeting people'. A prototype of the hardware was produced, but discretion prevailed, and it never went into production.

Then there was the cheekily-named Moogle, a mobile search engine that evolved into a classified advertising platform. Clockspeed were selling the odd game for international distribution, but the big breakthrough, the killer app, was proving elusive.

'We were just shooting up in the air, and hoping we hit a target,' recalls Gavin. Then, one day, he was working on a new mobile game, Faraway, which also had a monkey as its hero, when Herman walked in and told everyone to drop tools. There was a new project on the table, with a deadline of two weeks. An IM, or Instant Messenger, for mobile phones.

Gavin's first reaction was an inward groan. He felt 'quite pissed off' about having to abandon his monkey game, but Herman seemed gripped by an irresistible fervour, a sense that the earth was about to shift on its axis and everyone had better come along for the ride. They were cutting out all the complexity, all the gimmickry, and were going right back to the engine room of Alaya, to the simple idea of a community connected by messaging.

It was as if all the disconnected fragments, the shrapnel of ideas and opportunities and applications, were being drawn towards a common centre at last. A new story, a new mythology, was about to be written.

There was some sort of magic, too, in the alignment of the stars: the synchronicity of an impending drop in the price of mobile data in South Africa, rolling away the stumbling block that had kept the majority of users from accessing the internet on their phones. Data was about to tumble from R50 a megabyte, to R2 a megabyte. The time was right for a fast, cheap, easy-to-use IM. To be more precise, an MIM: a Mobile Instant Messenger.

Instant messaging, by the standards of the internet, is an antediluvian technology, going all the way back to the text-based Bulletin Board Systems that allowed early adopters to communicate over the phone line in the late 1980s. You would sign up to a BBS and dial-up on the modem attached to your computer, waiting for the overture of shrill electronic pulses and the satisfying hiss of static that confirmed your connection. Then you would log in, type your message in plain, blocky text on a black screen, wait for someone to type back, type your response, wait for someone else to type back, and so on.

The technical term for this mode of communication, conducted in real time over a data network, one person to one person or one to many, is synchronous chatting.

In South Africa, the biggest commercial BBS was Beltel, run by the state-owned telecommunications provider, Telkom.

For a subscription fee of just over R10 a month, R0.26 per successful log-in and port charges of R0.078 per minute, you could chat with other users on your 'Videotext' terminal, but that was just the start of it. You could also do your banking, pay your accounts, read the news and weather, go teleshopping, check share prices, send electronic mail, look up phone numbers and play online games. What you couldn't do was put it all in your pocket and take it with you.

So Herman wanted an IM that was mobile, an on-the-go app that would combine the quirky games with the social connections with the short messages with the trading of goods and services. It would be a mix of all these things, and more. And Herman wanted it in two weeks.

Because the primary capability of an IM is the exchange of messages, the working name for the app was Message Exchange IT, the IT doubling for Information Technology, or just plain 'it'. As in, 'I'll MX it to you.' MX-IT, for short. Pronounced 'Mixit'.

Before a team of software developers starts working on a

project, they are presented with a formal document outlining the objectives and functions of the application, the proposed user interface, and a walkthrough of the technical steps and processes required to bring the idea to fruition.

In this case, the goal was to develop a mobile 'client' application that would be able to access remote computers, or servers, over a network. The app would be built using J2ME, a variant of the popular Java programming platform, specially designed for mobile and wireless applications. Or as the functional specification put it:

MX-IT

Functional Specification

Version 1.0

1 November 2004

Owner: *Marnus Freeman*

Marketing: *Gerhard Gibbs*

Client Developer: *Gavin Marshall*

Back end: *Stephan Wright*

Description:

The MX-IT software will be divided into two components. An instant messaging and a forum component called MX-IT IM and MX-IT Forum.

The mobile J2ME client will be a single client with both functions in one. Users will connect to a hosted back-end server for both these services.

The user will register through the client interface and use his mobile number MSISDN as his user name.

The user will be able to add additional user names to interface with other services such as ICQ and MSN.

When the application is started a splash screen is provided

*with an additional screen providing branding opportu-
nities for distributors.*

*The client will also show a ticket service on the screen that
will be used for announcements and information.*

*The client will check for announcements on a regular basis
and present the user with a pop-up showing this. The
client will also check for new updates and instruct the
user to upload the new client.*

*The user will have free access to the service and will only
pay for the download of the client. This implies that the
client may not be distributed from phone to phone.*

At the helm of the project was Marnus Freeman, a contem-
porary of Herman's from the Computer Science Faculty at
Stellenbosch. Herman was the systems analyst, and Marnus
was a student, part of a group of renegades who relished the
challenge of hacking into the network, and seeing how long it
would take the admins to fix the breach. Later, he would join
Herman at Swist, and would help set up an internet service
provider for Vodacom, the mobile telco.

Marnus is the kind of guy you would want on your side on
a software development team, or for that matter, a rugby team.
He has the build of a prop forward, with a bullet-bald dome
to add a hint of speed to his bulk. But the most striking thing
about Marnus is the look in his eyes. They are a piercing blue,
almost laser-like in intensity, reflecting a retinal condition that
makes him highly sensitive to light. It's called tunnel vision.

This makes it difficult for him to see fine detail on a cir-
cuit board or a computer screen. And he chooses not to drive,
to avoid being a danger to other road users. 'But at night,' he
says, 'I can see better than most people.' One night, Marnus

was at home, chatting about the way cellphones have changed the way we chat, with his wife, Mariana, who once worked in nursing and is now an administrative manager at Mxit.

They wondered why there wasn't a better way of communicating by text on a mobile phone, a less intrusive, less expensive and less limiting way than SMS. And a more enjoyable way, maybe built around the concept of a multiplayer game. Without knowing it, they were defining the functional spec of the app that would become Mxit.

> They wondered why there wasn't a better way of communicating by text on a mobile phone, a less intrusive, less expensive and less limiting way than SMS.

Jaco was given the task of developing the client, the front end of the system, which would allow users to hook into the network, to send messages from phone to phone via the servers at Riesling House.

He took what devs call a 'blank canvas' approach, programming from scratch, rather than using the controls of the phone to guide the process. That puts a big burden on the team because they have to design with multiple cellular handsets in mind. Coding is architecture. You begin with a blueprint, a spec list of needs and wishes, and you work your way up to the sky.

Jaco used an open-source platform called Erlang Jabber Daemon, or ejabberd, as his foundation. Open source means free: free of charge, and free to customise, modify and tweak as you choose. But that freedom can come with a price.

The Erlang programming language is lightweight and powerful, but it was originally designed for PC networks, which are more stable and predictable than the mobile variety.

A mobile IM built on the platform may start feeling the strain when it reaches about 250 000 users signing on intermittently throughout the day, and flooding back en masse to check their messages after a network outage. When that happens, it can collapse under its own weight, bringing the whole system crashing to the ground. But who could see that far ahead?

It took Jaco three weeks to build the Mxit client, with other devs working on the back end and the server applications. By the end of November 2004, the system was ready for testing. Jaco signed up. He was User Number One. Beep.

It was after work. He was in his kitchen making supper. He checked his phone. It was a missed call from Grant Cleveland, a general manager at Clockspeed. That was how Mxit worked, in its early incarnation: if you wanted to chat with someone and they were offline, the server would send a discreet nudge to let them know there was a Mxit message waiting.

Jaco opened the app on his phone. There was Grant, his contact, his friend, his buddy. They began chatting, tapping short messages back and forth, like a twenty-first-century version of Alexander Graham Bell and Mr Watson.

'It was just idle chatter,' recalls Jaco. ' "What's up? So what are you doing? Nothing much. I'm going to make supper now" – that sort of thing. But I can remember very well, getting the feeling of, wow, this thing's actually going to work. The technology was cool, it was enjoyable to use, and you could chat for hours without having to worry about the cost. It all started to make sense to me.'

More than that, Jaco's undergraduate idealism, his desire to make a difference and be part of something that mattered, was about to be put to the test. In January 2005, Mxit was launched into the wild. Free instant messaging for the masses: a home-grown alternative to SMS, at last. But there was a catch.

Marnus had drawn up a 'business case' for Mxit. The company had to cover the costs of its internet operation, which were running at about R500 000 a month. So the model was that you could download the Mxit app for free, and you would get your messages for free, but you would pay a nominal monthly subscription fee of R15 to use the service.

Fifty cents a day. Not even the price of a cup of coffee. But the internet had created a new economy, based on the principle that you get what you don't pay for. You don't pay to use Google. You don't pay to read the news. You didn't pay, back then, to use the other MIMs on the market. So why would anyone want to pay to use Mxit?

A user named alchamy, on the MyBroadband online forum, summed up the mood in the marketplace:

> *www.mxit.co.za has launched a instant messaging service for cell phone via GPRS.*
>
> *Sounds great but they charge you R15 a month for access to the service: mad:*
>
> *What a joke, there are tons of free applications for MSN, AIM, Jabber, ICQ, etc.*
>
> *As far as im concerned anyone falling for mxit.co.za needs to learn about google.*

In a measured and cordial reply, Grant Cleveland pointed out that Mxit was more reliable than the other apps, supported more phones and transmitted between two to three times less data, thanks to its robust compression protocols. So it was good value for money. But even at Riesling House, there were doubts about the viability of the business model.

'You could pay your subscription via credit card,' says Gavin, 'and the first person who bought it, we looked at ourselves and thought, what idiot would do this?'

Mxit needed to attract some 30 000 subscription users to make the model work. When the number had reached a mere sixth of that, within four months, Marnus went back to the board and said, 'It's not working.'

Herman was in the process of buying out his partners at Swist. Money was tight. He took a decision that at the time may have seemed impulsive and unreasonable: 'Let's make it free and see if it flies.'

'And what if it doesn't?' asked Marnus.'Well then, we'll make a plan.'

In May 2005, Mxit went free. Liberated, untethered, it took on the momentum and meaning of something much more than an Instant Messaging application for the mobile phone. It became a political movement, rallying people to the cause with a principle of economic redistribution that came straight from the handbook of a legendary socialist hero. Robin Hood.

Mxit would be taking from the rich – the big telecommuni-

> Mxit took on the momentum and meaning of something much more than an Instant Messaging application. It became a political movement.

cations companies – and giving to the poor, the average citizen in the street, armed only with an average cellphone. You didn't need a smart or fancy phone to use Mxit. You could chat with your individual contacts, or enter the chat rooms and take part in a bigger conversation, and it would hardly cost you a cent.

As long as you could access the mobile internet, using GPRS or 3G technology, you would pay only for the data you used. To send an SMS with a maximum of 160 characters via your cellular service provider would cost you R0.75; for that amount, you could purchase more than 37 messages, each with a maximum of 2 000 characters, on Mxit. *Vive la révolution.*

The slogan at Riesling House, where Clockspeed would soon be rebranded as Mxit Lifestyle, was an unequivocal call to action: Join the Revolution. Gavin didn't like it. Revolutions just go round and round in circles, he felt. He suggested dropping the 'R'.

> The slogan at Riesling House, where Clockspeed would soon be rebranded as Mxit Lifestyle, was an unequivocal call to action: Join the Revolution.

He liked the idea of Mxit as an ever-evolving organism, a collection of cells, shifting, adjusting, colliding, connecting. An ecosystem, entire of itself. A social network. Join the Evolution.

'People come into the world,' says Gavin, 'and the world is a terrifying and scary place. Suddenly you're here, and you've got all this information funnelling into your body. That's why we need family, to give us structure and boundaries. Walls that protect us from the outside world. Social networks do the same thing. They're a means of survival.'

And all of this for R0.02 a message. But you couldn't just put that on a billboard and expect people to get the message. This was a revolution – no, wait, an evolution – and it called for a guerrilla campaign. Undercover rebels, pasting posters in forbidden places. Artists, poets and coders, handing out flyers and drawing up manifestos. A community of young activists taking to the streets to spread a promise of change and escape. This wasn't just a movement. This was an army: the New Mobile Generation. 'M-Agers', as Mxit called them – school-goers and students between the ages of 12 and 25. There were plenty of potential recruits in Mxit's university home town, and beyond that, in the townships and housing projects of the Cape Flats, where the network would soon gain its strongest foothold.

The guerrilla behind the campaign was a former BComm, law and economics student who had gone on to run a night-club in Stellenbosch, called Life. He sold the club, worked for an advertising agency for a while, and joined Mxit as a creative director, with a self-imposed mission to 'cause kak', but in a positive and creative way. His code-name, his 'nom-de-forum' at Mxit, was Traveller, which is what we'll call him here. Herman gave Traveller a once-off budget of R100 000, to spend as he wished, as long as he raised the user base to 125 000 within three months.

But Herman was already looking beyond that. The global market of GMS mobile users in 2005 was more than one-and-a-half billion, and Herman wanted to reach 2 per cent of them by 2008. That set him a target of some 30 million Mxit users within a mere three years. As it turned out, Mxit would have 7 million users, less than a quarter of the hoped-for total, by 2008. A classic case of shooting for the moon and landing among the stars.

For now, of that R100 000 creative budget, R5 000 would be spent on posters, and R95 000 on people: a graphic designer, a copywriter, and a social media director, hired for three months on contract. Nothing would be spent on print, radio or any form of conventional advertising.

The posters were comic strips, graphic novels, intricately evoking alien worlds, with elfin characters wandering through lush landscapes and cities that seemed to be made of living, organic matter. Somewhere in there was the virus of an idea, of Mxit as an alternate reality, a place where you could be your own hero, create your own story. The Mxit_Reality, it was called.

'The mobile phone allows you to have reach,' says Gavin. 'You become part of a complex and evolving ecosystem. Buddhism says everything is connected, everything comes from everything else. Nothing has existence in and of itself. There is no self. Form is emptiness, emptiness is form.'

By July 2005, there were 35 000 users. By September, there were 75 000. But the trouble with an alternate reality, is that sooner or later, it collides with its opposite.

Sitting on a scatter cushion in his office, he stops, frozen for a moment again. 'How any of this relates to Mxit,' he says, 'I'm not too sure now.'

But it worked. The campaign intrigued and captivated its target audience, turning something that had become everyday and mundane – instant messaging on a mobile phone – into the stuff of fantasy and myth. Mxit was reaching out, building a community, marketing itself with every message, every conversation.

By July 2005, there were 35 000 users. By September, there were 75 000. But the trouble with an alternate reality, is that sooner or later, it collides with its opposite. In the real world, Mxit was burning money. The original business model had been cast aside, in the hope that something better would come up. Herman was funding Mxit out of his own pocket.

He gathered his small team around the table at Riesling House, and he said, 'Listen, boys, we're out of money. We're going to have to shut this thing down.' It wasn't a threat. It wasn't an ultimatum. It was an incentive.

Somebody, somehow, was going to have to pull a rabbit out of a hat.

VIRTUAL WORLDS AND REAL MONEY

CHAPTER THREE

*Every high-tech start-up needs
a good business model to turn a
big idea into profit. In the case
of Mxit, it was the coining of
a currency and the creation of
new worlds of communal chat.*

A WHITEBOARD is an ocean of the imagination. Calm and glossy on the surface, it harbours hidden depths of insight, cresting on waves of inspiration that can carry you all the way to the shore. In the squeaky scribble of the Magic Marker, in the mind map of arrows and circles and numbers plucked from the air, you can see the future being born.

Gavin Marshall, head of innovation, conjurer-in-chief, walked into the boardroom at Mxit, surfing on an idea that could, just maybe, save the company. A strategy for revenue generation. A product development opportunity. A business model. And all it would cost was – pluck – one cent. A single South African cent.

It was the winter of 2005. Mxit, with its more than 70 000 users, and hundreds signing up every day, had figured out almost everything – the system, the technology, the spreading of buzz in the marketplace. What it hadn't figured out was how to make money.

The R15-a-month subscription model that had launched Mxit as an almost free mobile instant messaging service was lying on the ocean floor, wrecked. The advertising model, based on the selling of space on the splash screen that greets you when you log on to Mxit, held some promise, but not enough to keep the company afloat.

The test case had been a campaign, a few months back, that looked like an ad from the classifieds, the personal section, clipped from the paper and ringed in red lipstick. 'Looking for love', it said. 'Black, Slim & Sexy, seeks partner, GSOH'. Good Sense Of Humour. And just below that, a tiny picture of a Samsung mobile phone, black, slim and sexy.

The ad generated 20 000 'unique splash screen views' and 14 000 hits on a micro-site in 72 hours. Here was a new advertising medium, a

> Mxit had figured out almost everything, except how to make money.

mini-billboard for the mobile generation, the 12- to 25-year-olds with disposable pocket money or income of their own. But for now, Mxit was coasting on the charity of its founder. Herman was running out of patience and venture capital.

Come on, boys, he had said. We need to come up with something. Gavin was about to pitch his big idea to an informal audience of two. There was Grant Cleveland, the general manager, and next to him, the creative director, the man who had most come to personify the culture of the 'underground rebel network', with its carefully crafted fusion of the real and the fantastic. He called it the Mxit_Reality, and he called himself Traveller.

On Mxit, you can be anonymous, a character of your choosing, unburdened by the perceptions and expectations of other people. So we will grant his petition, and refer to him here only by the name he uses on the Mxit online forum. This is the prerogative of the creative director.

A sinewy figure, with black hair and a chiselled jawline, he projects the spring-tight physical energy of a rock climber, poised between footholds. As it turns out, mountaineering is one of his great passions.

He talks of a peak just outside Stellenbosch, a deceptively easy climb, until you reach a crest that reveals the true summit jutting high above you. The Devil's Tooth. And then you stand on the top, and the mountain flares below you, and it feels as if you are wearing it like a dress. He met his wife on a climb. She is a strong and determined climber. 'I thought she was a man at first,' he says, 'because of all the climbing gear.'

> On Mxit, you can be anonymous, a character of your choosing, unburdened by the perceptions and expectations of other people.

Now in his early thirties, he seems to move through the world of Mxit like a spirit, defining his own terms, plotting his own journey, guided by the notion that we are all on a path of discovery that will eventually lead us back to ourselves. Hence 'Traveller'.

Once he took a two-year break from work, abandoning his worldly possessions, as a 'contained social experiment' to see what it would be like to depend completely on the kindness of friends and strangers. 'Our society programmes us to be ambitious, to be successful,' he says. 'I thought if I could break that down, I could become something I'd like to be, rather than what society expects me to become.'

Years later, when he was working for Mxit, he had a raging

argument with Herman over the direction the company was taking, and the mechanics of a staff trust scheme. He didn't want to work for an organisation where the dream was to make one person rich. So he left and went to the Amazon to learn from the shamans. Always searching. Always questing.

It was in 2004, as an avid gamer, that he first applied for a job as general manager at Clockspeed, even though his real obsession was the desktop multiplayer strategy game, rather than the handheld mobile diversion.

His game of choice was World of Warcraft, a dark and absorbing series of quests through kingdoms populated by dragons and orcs and demigods and zombies and trolls and scourges and the occasional human. He decided he didn't want to be a general manager after all.

A year later, just after Mxit had been invented, Herman called and offered him a different position, as a creator, an agitator, an architect of the impossible. Officially, creative director. He looked at Mxit – there were 65 users when he joined the company – and he found the product itself to be an 'ugly, complicated and functional thing'.

In the absence of visual appeal, he felt it needed an aura, a narrative that would transcend its simple utility as a mobile instant messenger. But it wasn't the system that grabbed him. It was the radical, utopian ideal at the heart and soul of Mxit. Free telecommunication for all.

In practice, virtually free, because you were still paying micro-amounts for your airtime and data. But free to choose, free to belong and free from the hefty per-message costs of SMS on the cellular networks. He applied his BComm mind to the challenge: how were you going to make money from a product you were virtually giving away?

Clearly, you were going to have to make it on the peripheries, rather than at the core, where your sacred contract with the user was instant messaging for next-to-nothing. Gavin had the answer. Chat. Not one-to-one, but one-to-many.

It's funny how sometimes the best ideas are those you bin as unworkable to begin with. Then, out of intuition or desperation – in this case, Gavin had just been looking for a way to hang on to his job – you haul out the idea, un-crumple it, dust it off, and re-examine it in the light of a change in circumstance.

Mxit had been designed as a pure instant messaging client, with no facility for conferencing or communal chatting in a designated virtual chat room, over an internet connection.

The chief obstacle had been the physical medium: the small screen of a mobile phone makes it difficult to follow a series of synchronous messages, back and forth between multiple participants. But what if you were to limit the number of people in a chat room to, say, seven? And what if you were to charge them, say, one cent per message?

> But what if you were to limit the number of people in a chat room to, say, seven? And what if you were to charge them, say, one cent per message?

The payment mechanism already existed, in the form of the premium-rated SMS system that had been built into the multiplayer mobile games at Clockspeed, right from the start. Even better, you could come up with a form of 'virtual currency', unique to Mxit, that would add an element of playfulness and swagger to the transaction. You wouldn't be splashing out rand and cents, you would be spending … Mxit money.

Herman breathed a 'visible sigh of relief' when he heard the idea, recalls Gavin, and he gave the go-ahead for the chat room software to be written. It took Gavin two weeks.

In the other-world of the Mxit_Reality, where you were invited to shed your identity and embark on a journey of fantasy and discovery, he imagined the chat rooms as distinct spaces where you could hang out, make friends, talk about whatever was on your mind, or trade inventive insults and banter just for the fun of it.

Each chat room would be part of a bigger chat zone, to increase the array of choices and make room for as many chatters as possible, with a maximum of seven per room. (He later upped the number to ten, just the right balance between cosy and crowded.)

Because the origins of Mxit could be traced to the rocketship that you would steer by SMS in the first Clockspeed mobile game, Alaya, it was perhaps inevitable that the first Mxit chat zone would be moored among the stars.

It was called AlphaCom, and an artist visualised it as a giant floating hub, titanium in colour, with sixteen domed berths radiating from a central core, the whole space station surrounded by a glowing atomic force field. It looked, from the top down, like the classic representation of a computer network, a circle of interconnected nodes positioned in a virtual space.

More down to earth was the second Mxit chat zone, Boondocks. It was a nightclub, straight out of *Saturday Night Fever*. There was a dance floor made up of multicoloured glass squares, a lounge and bar, a courtyard with umbrella tables, a DJ slot, a dancing cage, a chill room and a bathroom.

Or you could rig the sails of your wooden ship and set out for Fantasy Island, with its steamy jungle, smouldering volcano, luxury hotel and lagoon.

All these were provinces of the mind, artists' impressions of what a chat zone would look like if you could cross the digital divide and be there with your chat buddies in person. But they were real enough for you to suspend your disbelief, just like you do in the movies.

'COOL!!!!!!!!!!!!!!!!!!!!!!!!!' exclaimed a user on the Mxit online forum, where the chat zones were announced and put on display. 'I dig I dig!!!! Man I BOW down to you guys!!!' said another. And someone else, speechless, could only issue the emoticon of the gasp-face: :o

> You can't see or touch Mxit Moola, but it is as good as cash on the network. Better, even, because you pay up front for the services you are planning to use.

That, too, would be the reaction at Mxit when the Moola started rolling in. Moola. The name of the Mxit currency was coined by Johnny Vergeer, a colleague of Herman's at his principal company, Swist. It was borrowed from a slang term of indeterminate origin, meaning money. You can't see or touch Mxit Moola, but it is as good as cash on the network. Better, even, because you pay up front for the services you are planning to use.

The idea was that you would send a premium-rated SMS, costing R2, R5, or R15, to buy the Moola you needed to send messages in the chat rooms. The official exchange rate was pegged at 100 Moola = R1, so R2 would get you 200 messages, at 1 cent a message. Small money. But it quickly turned into Big Money.

In the first month of the chat rooms going live on Mxit, says Gavin, they generated R20 000 for the company. In the second month, R50 000. Then R100 000. Then, in November, 2005, a small change in the business model.

One cent, or one Moola a message, had sounded like a good idea at the time. But the chat rooms were busy: so busy that they needed a traffic management system to keep the conversation flowing. If you were in a room, and you had nothing to say, you would be 'nudged' after three minutes by an automatic message: '*You are rather quiet in this chat room.*'

If you carried on being rather quiet for another four minutes, you would be booted out to let someone else have a chance. In the real world, no one will judge you too harshly if your contribution to the social discourse is limited to an occasional nod or shrug. Maybe you're just a good listener.

But in the virtual world of the online chat room, where body language is invisible, that sort of behaviour is frowned upon. It's called lurking. It inhibits communal engagement and repartee, the core principles of good online chat. And of course, it's not good for the economy.

So the number crunchers at Mxit, having established the basis of pay-to-chat, wondered whether users would be prepared to pay a little more. Just one little Moola more. Which was a hundred per cent more, if you wanted to get technical about it, but still, as this announcement on the forum suggested, good value for Moola:

Double the price!!!

Can you believe it, we are going to increase the cost per message from 1 moola to .... 2 moolas. :D Well, up until now, most Mxit Chatroom users would have referred to 1 moola as 'nothing', which is true, as it is equal to 1

> *cent – which is nothing. So if you think about it, 2 x*
> *nothing = nothing. So you still pay the proverbial 'noth-*
> *ing' per message in the chatrooms. :twisted:*
> *I confused myself there for a moment.... Anyway, as of next*
> *week (Monday) you will pay 2 moolas per message as*
> *opposed to 1 moola. This is to cover our very much*
> *HUGE... internet bill.*
> *We hope (and know) you will understand. Regards :wink:*

Naturally, there were rumblings – 'you guys are getting greedy ... you're probably pulling in +- 50 grand a day!' – but for the most part, the market shrugged, nodded and carried on chatting. 'It was astounding,' says Gavin. 'Within six months, we were making a hundred thousand rand a day in the chat rooms, at two cents a message.'

By September, 2006, a year and four months after the launch of the free instant messaging service, Mxit had broken even. Today, the chat rooms still account for 30 per cent of revenue, and the cost per message is still 2 Moola. It must have been tempting, over the years, to double that figure again, but the business case for leaving it be is compelling, as Gavin explains.

The data costs per message amount to 0.0002 cents, or one five-thousandths of a rand. If you factor in all the operating costs, including server costs, electricity, salaries, share of revenue with the cellular networks, everything, you're still making a profit of 1.19 cents on every message sent for 2 cents. It's micro-money, but it adds up. And the Moola isn't just for messaging.

The chat zones on Mxit – 65 at last count, each subdivided into a series of rooms – are categorised according to age, location, fantasy and special interest. So you can chat about life in

Jozi, Durban or Cape Town, meet your fellow thirty-somethings in 30 Something, battle it out with words and emoticons in The Battleground, and embark on a hero's quest in Dark Mountain. Or a villain's quest.

The chat rooms are mirrors of the real world because they are peopled by real people, masked by their nicknames, but equipped with real emotions, real motives, the real power to charm or amuse or provoke or confront. Sometimes they mirror the dark side of human nature, and that's where the .rat command comes in. Dot rat.

If Mxit is a country, and the chat zones and chat rooms are its cities and suburbs, then the .rat command is the signal of the neighbourhood watch on patrol. The chat rooms are monitored at random by the support team at Mxit headquarters, and by trained part-time moderators or 'mods' across South Africa.

Most of the time, they lurk unseen in the rooms, using a special 'hide presence' command, which allows them to keep an eye open for profanity, abuse, hate speech, spamming, offensive or inappropriate nicknames and 'sexual misconduct'.

In the support centre at Mxit, where the team sits at computers around an oblong table, Ryan Miller, senior chat zone administrator, leans across to look at a query that has just come in from a mod. She wants advice on how to handle an apparent crossing of the line by a user on a Rainbow Cruise chat room. His suspected transgression: he entered the room and typed a two-word message, 'I'm horny'.

Ryan asks the admin to mail the mod. 'That's a BFL,' he says. A ban for life, for sexual misconduct. The user has the right to appeal, but if you use the chat rooms, you're supposed to know the rules, and you're supposed to know that somebody might be keeping watch. But users are granted that power themselves, in the form of the .rat.

You use it, by typing the command in your chat room window, to 'rat' on a user who you believe is being abusive, crude, or hurtful, or who you suspect of being under or over age in an age-restricted room. A teen in a 30 Something room, for instance, or vice versa.

The command sends the most recent 30 lines of the conversation to the Mxit admins, and they decide on the merits of the case. Up to 500 .rats a day are sent to Mxit, most commonly for sexual misconduct or age transgressions. So what's to stop a chat room user from deploying the .rat command with malicious intent, as a form of harassment or abuse in itself? Ah.

'The idea with the dot rat command was to give the community a means of moderating itself,' says Gavin. 'But we soon had a lot of people complaining that they were being banned from the chat rooms, and they were doing nothing wrong. We did some investigating, and we found that guys from the Cape Flats were using dot rat as a weapon.

> Send someone a virtual flower for five messages. Send them a coffee or an ice-cream for three messages. Give them a hug or a slap or a tickle for two messages.

They would lie in wait in the chat rooms, and ambush users by dot ratting them out of the room.'

That taught Gavin two things. One, the .rat system wasn't working. Two, sometimes, in a chat room, you need a weapon. So Gavin built extra dot commands into the system, including a .bomb for those with a short fuse, and a range of more socially acceptable accessories for the better angels of our nature.

You are free to bestow these items on anyone who pleases or displeases you, but the catch is, they aren't free to bestow.

As a disincentive to malicious .ratting, each .rat will set you back five messages, or 10 Moola, and you can also use the currency to add value to your interactions in the chat room.

Send someone a virtual flower for five messages. Send them a coffee or an ice-cream for three messages. Give them a hug or a slap or a tickle for two messages. If you really need to know, send them a .doyoulikeme, for two messages. Or send a .amihere, for only one message, to check whether you're still in the chat room. It's easy to see why some visitors would feel the need for an existential affirmation.

These are worlds unto themselves, where people who seem to know each other well communicate in cryptic staccato bursts of text, punctuated with pictograms to convey emotion:

> **My_name_is_skrillex:** *You stupid dog! Yoy make me look bad! -puts on scary mask- ooooga boooga booga!*
>
> **AcId8-oDiPt8-oPuPPy:** *O.o -lifts paw- if I had fingers I'd extend a finger*
>
> **TheAshtrayGirl:** *I'd love to stay. But i can't. Goodbye* 🙂
>
> **TheAshtrayGirl has left the room**
>
> **Princess(*)Fairy has entered the room**
>
> **Hullaboo has entered the room**
>
> **AcId8-oDiPt8-oPuPPy:** *I haz one of ichigo's hollow masks* 🙂
>
> **Princess(*)Fairy:** *hey hey*
>
> **((Zangetsu>>** 😗
>
> **Hullaboo:** *Good evening, everyone*
>
> **My_name_is_skrillex:** *I want it!!!!*
>
> **Princess(*)Fairy:** *my neighbor sacred me with scream mask. i threw rocks. i gave. he peeping tom. so sweet*

> *AcId8-oDiPt8-oPuPPy:* 😮 *and I'm recreating Lights Death*
>   *note book o.O*
> *You are rather quiet in this chat room...*
> **$AMURA![UAD]DRAGON has entered the room**
> **Monkey I D I Luffy has entered the room**
> ***AcId8-oDiPt8-oPuPPy:** o.O me.. Not.. Understandingz*

That's from a night in Dark Mountain, the role-play chat zone where you can be anyone or anything you want to be. It's a non-aslr zone, which means you're not allowed to ask the 'age sex location race' question, which is the standard getting-to-know-you opener in the flirtier rooms. But you learn the social niceties, and you get to know the places and people where you feel most at home.

For Kay, a thirty-something operations manager for a financial institution, that zone was Cocktail, where at first she felt overwhelmed and out of place amidst the babble of the 'regulars clique'. She was used to one-on-one chatting on Mxit, and it took a while for her to feel at ease in the room, to be accepted as a guest who could bring something to the party.

In the real world, she says she is a klutz, a sore loser, a recluse or a social butterfly depending on her mood. But in the chat room, she enjoyed being able to 'put reality on hold, to be entertained and have a social experience without having to dress up for it.'

She discovered she had a flair for chatting, for ushering people into the conversation, for listening and learning and understanding something about the way people act when they're free to be someone other than themselves. So she applied to be a chat zone moderator.

After enduring a three-week online 'boot camp' to test her mental strength and ability to deal with the unsavoury side of chat room culture, she was appointed to a squad of mods called the Mxit Magi. Her task: 'to rid the chat rooms of unruly characters'. Well, that's how she saw it at first.

Now she would rather try to persuade people to change their ways, to be better Mxiteers, at least where the lesser offences of profanity and 'virtual abuse' are concerned.

She wants the users to see the mods as people just like them: 'We chat, we have fun, we make them laugh,' she says. 'We show them that moderators are indeed human.'

One of her fellow mods is a bartender by the nickname of vampish17, who lives in Cape Town and who hopes one day to qualify as a teacher or lecturer. She spends about five hours a day chatting and moderating on Mxit, mostly in the day because she works at night.

She chats to her mother and some close friends, and she browses the XChange classified ads too: 'I found my kitten there.' But mostly, she uses Mxit as a wall, a barrier against unwanted contact.

'Sounds silly, but I'm shelled up that way,' she says. 'I love my privacy. I don't allow people to really get to know me in person unless they try continuously for a long period of time.'

She spends a couple of hours a day moderating, answering questions from users, changing inappropriate nicknames, banning the odd user for sexual misconduct or profanity. Being a mod has taught her patience, diplomacy and discretion, and if the need arises, she now knows how to defuse a situation in real life too.

Then again, if you spend a little time in the chat rooms, getting to know the cast of curious characters who drift in and out, you'll find that they're also a bit like real life. All you have to do is use your imagination.

Awakening the Dragon

CHAPTER FOUR

*The long view on starting up,
starting over and rebuilding
the culture of a company.*

O<small>N A CLEAR</small> day in Stellenbosch, you can see the sentinels of stone that encircle the town, like a Council of Wizards watching over us as we plot and toil in the valley. Sometimes, we watch back.

To live in Stellenbosch is to take the long view, and the sight of the mountains never fails to fill me with awe and heighten my sense of perspective.

I think of how, millions of years ago, they rose from the earth in fire and chaos, shaped into twists and folds by the collision of continents. And today they're just part of the scenery, watchtowers of the Winelands, the brooding beacons we navigate by.

But there is one range that captivates me more than the others, and that is Groot Drakenstein.

It lies to the south of Stellenbosch, and in winter its peaks are lightly covered with snow, giving it an Alpine aspect when viewed from afar. To me, the range has always looked like the spine of a dragon in repose, an image I'll admit was planted in my mind by the name alone. Groot Drakenstein. Great Dragon Stone.

It pleased me that the early colonists, in the seventeenth century, had been able to see a fire-breathing, serpentine beast lurking in the rifts of shale and sandstone. But then I discovered that the actual inspiration was a man with the imposing name of Hendrik Adriaan van Rheede tot Drakenstein, a military commander and colonial administrator in the service of the Dutch East India Company. Ah, well. It still looks like a sleeping dragon to me.

I've always been something of a drakonophile – a person with an inordinate fondness for dragons – seduced by their glowing eyes, their leathery wings, their roar of flame and by the secret knowledge that all of this is just a test of faith for their ability to bestow good fortune on true believers.

Dragons, in my experience, are a fearsome but misunderstood species. I never could figure out why knights in shining armour had to set forth to slay dragons with their swords. A true hero would rouse the dragon from its lair, edge a little closer and say, 'Hey, can you show me how to do that fire-breathing thing?' And quite possibly, get scorched in the process.

But it's worth nothing that one of the words at the root of the English 'dragon' is a Greek word, *derkesthai*. It means to see things clearly. It means to have a vision. And so, on 1 November 2011, I pulled up at Technopark in Stellenbosch, at the start of my quest to awaken the dragon.

Technopark is a cluster of office blocks overlooking a wine estate and the mountains. At that stage, it was one of four sites housing the Mxit operation in and around the town, physically subdividing its technical, support, sales and administrative functions.

There may have been sound reasons for this, logistically

and financially, but it struck me as a symptom of a fragmented and cumbersome bureaucracy. Worse than that, it reinforced my preconception that the ethos of Mxit would prove to be characterised by the thing I fear most in any organisation. Not incompetence or over-ambition, but mediocrity. The luke-warmth of the middle road, the easy slide into the self-belief that while you may not be the best in the world at what you do, at least you're not the worst.

This is not the 'We Try Harder' of the Avis ad, where you are driven by a vision of yourself as a looming, nagging presence in somebody's rear-view mirror, right up to the point where you cheerfully overtake them. (Can anyone even remember who No 1 was supposed to be in that famous ad? I certainly can't.)

I'm talking about the 'We Hardly Try', about the acceptance of your perceived limitations as an immutable law, coupled with the twin evil of relativism.

'We're not the fastest, but we're not the slowest.' 'We don't have the best service, but we're better than most.' 'We're not the market leaders, but at least we're not last.' 'I'm not the best dad in the world, but I'm not the worst.'

Me, I'm not the best auditor in the world. I don't think I'm the worst, but I wouldn't really want to put any money on that. So I don't audit. Simple.

If you're not excelling at what you're doing, change what you're doing. How will you know whether you're doing something you're good at? You'll be loving every minute. Or at least, you'll love having done it, once it's done.

Excellence doesn't mean perfection. It means striving to do the best job you possibly can, in everything you do. It means being open to criticism and fixing mistakes. It means learning. Most importantly, it means chasing absolute goals and not measuring yourself against anyone else.

It pains me, as a South African, to hear how we must aspire to be 'world class', as if we are somehow shut off from the world and not entitled to be a part of it. In sport, science, technology, the arts, commerce, industry and leadership, we've proved often enough that the rest of the world would do well to aspire to our class.

And in Mxit, I saw a working model of that: a simple, nimble tool that allows people to communicate via their mobile phones for next to nothing, while also acting as a platform for social change, upliftment and economic freedom. An Africa-class technology, as good as anything else in the world.

> It pains me, as a South African, to hear how we must aspire to be 'world class', as if we are somehow shut off from the world and not entitled to be a part of it.

And yet, for all my bullishness, for all the due diligence, for all the conversations I'd enjoyed with some of the key players at the company, I was still expecting the culture at Mxit to be slow, bureaucratic and conducive to mediocrity. How wildly wrong I turned out to be.

I'm happy to say that at Mxit, I have met some of the smartest, coolest, most switched-on people I've encountered anywhere.

I have met coders, engineers, architects of the imagination. I've met artists, philosophers, designers, mythologists, jesters, adventurers, mind-bending magicians. I've met deep thinkers, fast talkers, idealists and fantasists. I've even met a couple of really good chartered accountants.

And then, in that first week of November, just after I learnt everyone's names and heard their stories, I had to let 15 per cent of the workforce go.

I felt like the Grim Reaper, sweeping in with my scythe, downsizing, rationalising, economising. There's no way to sugarcoat it, to make the cut of the blade any less painful. I just wanted to get it over and done with as quickly and as fairly as possible, to avoid the lingering uncertainty and unease that accompanies a change in the company regime. And I wanted to do it myself.

I wanted to be able to look people in the eye and take whatever flak was coming my way. And boy, did I get it. Here are a couple of things I've learnt about How to Win Friends and Retrench People.

> I wanted to be able to look people in the eye and take whatever flak was coming my way. And boy, did I get it. Here are a couple of things I've learnt about How to Win Friends and Retrench People.

Firstly, there seems to be a positive correlation between how much people earn in an organisation, how little real value they add and how aggressively they react when you tell them you sadly have to let them go.

Although they have the biggest safety nets, the highest paid are the ones that resort to verbal violence and threaten to sue. I think this is a function of an unconscious awareness of incompetence and a realisation that they will struggle to find another sucker to pay them such a salary.

The injustice of the situation is that they always get paid bigger severance packages than those who are paid less. My advice: suck up your ego, and pay whatever you have to pay to make the poison go away. Let fate take care of justice.

The nicest people generally don't fight. There are tears and sadness, but the emotion is more disappointment than anger. The most heart-breaking line is 'I'm so sorry I can't join you on this adventure'.

You are destroying a person's life and that person can be so humble that they actually sincerely wish you luck. These moments made me cry. My next piece of advice is, to do it fast.

Starting on the Monday, nineteen people were retrenched; all the severance letters were signed by Thursday. Morale was dismal for that week, but once the process was completed and an email had gone out saying no more cuts, everyone could settle down.

The secret to speed is: be honest, follow the letter of the law and don't quibble on settlements. The most important thing to remember, for me at least, is to respect a person's right to dignity.

The retrenchments at Mxit were not a reflection of the competence or integrity of the affected individuals, so they were able to walk away with their self-respect intact. And that is a commodity worth more than all the money in the world.

But I also learnt something about the importance of a human resources department in a small to mid-sized company. And that is why, today, we don't have a human resources department at Mxit.

I don't believe there is anything human resources can do that humans can't be resourceful enough to figure out on their own. Of course there are rules, systems, processes, practices, best ways of working. But these things should be inherent and instinctive, or at least absorbed by osmosis. Likewise, there used to be a unit at Mxit called 'business intelligence'. The idea was that they would gather intelligence from the marketplace, buzz from the industry, input from consumers, news and insights on trends and developments. Then they would feed all of this to the rest of the business as a tool for strategic planning. Well, you know what happened next.

Queues of people, suspended in limbo, waiting to get their business intelligence reports before they could make a move. The job of thinking critically, scouting for data, sharing, filtering, analysing and interpreting information had been outsourced to an internal department. So: delete. Delete HR, delete business intelligence. Delete, delete, delete.

People were flying blind, bumping into each other, grasping for the handholds of the status just gone quo. Then, slowly, they began to find their wings.

For a few days, there was chaos and confusion, the natural consequence of an organisation evolving from disorganisation to reorganisation. People were flying blind, bumping into each other, grasping for the handholds of the status just gone quo. Then, slowly, they began to find their wings.

The funny thing is, the model for this mode of working – taking ownership and taking flight – was Mxit itself.

In the early days of the network, when there was no call centre and few resources to assist users with set-up and technical problems, the small crew at Mxit started an online support forum that swiftly outgrew its original terms of reference. It became a database of useful knowledge, a platform for random

you change the culture. New office, new desks, new coffee, new restaurants. New company.

The final step, after the retrenchments and the move to La Gratitude, was to create a culture of trust, not fear.

There used to be a tradition at Mxit that whenever the coders started work on a new project, they would have to sign letters of resignation, which their CEO would file in his drawer. If you messed up or missed a deadline, your resignation would take immediate effect.

I suppose it kept people on their toes. Management by paranoia can be an effective way of getting things done, at least in the short term. The same philosophy can be put in place to sharpen everyone's focus on the job at hand. So YouTube, Facebook and Twitter had been blocked at Mxit.

I could see the argument against bandwidth abuse and distraction. At the same time, we're a tech company. We're a social network. How can you stop people from socially networking? Why would you want to? I lifted the blockade.

In its previous life, Mxit was a place of secrets. Whether you were staff, business partners or the media, everything was on a need-to-know basis. Sometimes, the information was contradictory, and the air would be thick with rumour.

Although I understand the rationale, the insidious side effect was to plant seeds of distrust. Why are you hiding if you have nothing to hide?

We started out with the small things. A new tradition: Thursday night drinks at the restaurant next door, with an open bar tab. Okay, maybe that wasn't such a great idea. A R21 000 tab on the first night, and guys were ordering multiple pizzas to take back home.

In my old life, I would have banned them, and stopped

the tab. Now I say, hey guys, I hope your families enjoyed the pizza. And I rank them, with a scoreboard in the office, according to how much pizza they took home. I just want to shine a light – on the good and the bad things.

I'm not a libertarian. I believe you need government, not to tell you what to do, but to provide the structure and systems that make things work. The community will take care of itself. Trust.

One day, I got fed up with all the niggly paperwork, so I changed the procurement policy at Mxit. Now everyone in the organisation, at any level, can sign cheques for up to R10 000 without approval.

People were saying, but what if 100 people go out tomorrow and spend R10 000 on junk for themselves? Well then, we lose R1 million, and we discover that 100 people are crooks. This hasn't happened. Trust.

The initial reaction to my brutal transparency policy was mixed. Turns out people want the nice truths, not all the truths. But, as the weeks have gone by, the culture has changed.

We have monthly presentations of the company's management accounts, informing staff of our exact cash position. We're honest with the media about user numbers and the user experience.

We have fewer meetings. Fewer group emails. We distribute Exco minutes to anyone in the company who wants to read them. Except we don't have an Exco.

We have a Council of Wizards. There are seven of us, all interchangeable, all replaceable. We sit around the table, drinking our coffee, and talking about dragons. I suppose this is my fault.

But I like to think we're building something here, the exo-skeleton of a big beast that will one day breathe fire and take flight over the mountains, over Groot Drakenstein, over Africa and over the seas. On a clear day in Stellenbosch, you can see the future.

THE MILLIONAIRES OF MOOLA

CHAPTER FIVE

*Clicks, taps and connections.
A cashless, cardless currency.
A hierarchy of Citizens,
Freemen, Traders and
Rockstars. A glimpse inside the
mobinomics that makes the
world of Mxit go round.*

COFFEE. Coffee, coffee, coffee. It was a crisp summer morning, and I was strolling down Church Street, when I heard the roasted arabica calling my name. I zigzagged from my route, crossed the road and pulled into the red-and-white way station on the corner of Bird.

The Vida was pumping. Samba music, the clinking of spoons, people talking and laughing. Stellenbosch is a small town. I waved, shook some hands and swapped some greetings and chit-chat over the noise. Then I went up to the counter to place my order. Large cappuccino to go, please.

The barista grinned as he rang it up. He knew the drill by now. 'Six four nine, one eight six one,' I said, looking at my phone. He punched the code into his point-of-sale machine. Match.

The first couple of times you do this – use a telecommunications device pre-loaded with invisible money to purchase a product or service over the air – it feels like magic. Then it feels like common sense: why shouldn't you use your mobile phone, the one device you always carry on your person, as an instrument of everyday financial negotiation?

Then it becomes a habit, ingrained into your lifestyle, wired into your muscle memory, and you don't even think about it until you run out of battery or the machine goes offline.

Even in the twenty-first century, cash and plastic make the world go round, but I still get a kick out of tapping my iPhone to pay for a shot of caffeine. It's all part of the Mxit Wallet, the digital trade and payment system that puts the working capital into the Republic of Mxit.

The goal is a cashless, cardless society, where your handset itself becomes a financial instrument, facilitating a connection between parties in a currency of taps or clicks. Once, the cellular phone was a device that allowed people to talk; now it's a device that allows money to talk.

Three clicks, and you've got a deal, sealed by a numeric code that puts you and the seller on the same wavelength, whether you're buying a coffee or paying for a taxi ride or tipping a car guard when you've run out of coins. Almost everyone carries a phone. Almost every

> Once, the cellular phone was a device that allowed people to talk; now it's a device that allows money to talk.

phone can carry money. It's still a goal, but it's not a dream.

When World of Avatar moved into the La Gratitude building in Dorp Street, all of us under the same roof, it didn't take long for people to point out the glaring omission in the bouquet of on-site facilities. A company canteen. Even coders cannot survive on coffee alone.

Then again, this is Stellenbosch: a student town, in the heart of the Winelands. There are fast-food joints, coffee shops, bistros, pubs, pizzerias and fancy restaurants all over the place. So we decided to try a little experiment. We gave everyone in the building a lunch allowance of R43 a day, to be spent at a selection of ten restaurants in the vicinity.

The catch was that the money was loaded into your Mxit Wallet, and you had to pay by phone. Actually, it wasn't a catch. It was a perk. We got a bunch of beta testers to try the system out, and for their trouble, they got food.

In the first month of the Mxit Staff Lunch Campaign, just over R115 000 was redeemed by 174 World of Avatar employees at the ten participating establishments. There were no major glitches.

Staff and merchants were happy, and the ease of use of the Wallet system – add a contact on Mxit, generate a seven-digit 'wiCode', share it with the merchant, authorise the payment – was confirmed over and over again in the real world.

In the real world, particularly on our continent, most people do not have access to debit or credit cards. So a phone with its own built-in wallet can make a very real difference to the way people manage and control their spending. Then, as I trawled the data, I noticed something interesting.

The biggest slice of the Mixt staff lunch money, more than 35 per cent, was being spent at the Kauai outlet, a health-food franchise known for its salads, wraps, and fruit smoothies. But Kauai is several blocks away from our office, a much longer stretch than the cake shop, pizza place or Mugg & Bean.

I thought this was an impressive testament to the health-consciousness of our workforce, until someone pointed out that the menu at Kauai was cheaper than most of the other options. So people were prepared to trek the extra mile to get more lunch for their buck. Simple mobinomics.

There is something about mobile technology that encourages you to think small, to rationalise and ration your spending, all the more so if the commodity in question is the very oxygen that enables you to use your phone in the first place. Cellular airtime, paid for in advance, allotted in increments, has been the real impetus for the growth of mobile technology in Africa, and it brings to mind the parable of the Indian sugar executive.

The story goes that a big consumer-goods company was looking for ways to boost sales of sugar, which was being sold in a standard 1 kg packet. An executive said, let's split it into two 500 g packets. So they did, and revenue shot up by 20 per cent.

Then they split the 500 g packet into two 250 g packets. Revenue shot up again, by 20 per cent. And so on, until the sugar was being sold in tiny sachets. Net result: double the revenue, and double the sugar sold. Sweet.

In nature, we know this process of splitting and resplitting as cellular division, and it is the basis of life itself. In the cellular industry, we call it prepaid. Drilling down the quantity of air you can purchase, from per month to per minute to per second.

In South Africa, prepaid airtime was launched by the Vodacom network in 1996, and it immediately multiplied the market of potential users by millions, eliminating the burden of credit vetting and the fixed two-year contract, which had kept the majority of the populace on the wrong side of the cellular divide.

Today, more than 85 per cent of South African mobile customers use prepaid airtime. They get the benefit of customisable affordability, paying for their voice and data calls in discrete segments, while the networks get the benefit of higher per-minute rates and exponential boosts in revenue. Simple mobinomics.

If you have limited cash, you cannot commit R100 at the beginning of the month for airtime. You would rather pay R2 per minute as the month progresses, just in case you are confronted on day ten by an emergency requiring R50. If you had committed the full R100 at the beginning of the month, you would be in trouble.

But even if your airtime evaporates completely, you can still use your phone to send a message asking someone to call you back. One of the big divides in the South African economy is between people who know how to send a 'please call me', and people who've never needed to. But if you do need to, it can be a lifesaver. Here's the equation. Limited money = limited flexibility = demand for more control of daily spending commitments = bite-size chunks of airtime = prepaid. The same model applies to Mxit, where you stock up on airtime as you need it, by sending a premium-rated SMS that is converted to 'Moola', the name given to the Mxit currency. 100 Moola equals one South African rand.

So a two-rand SMS gets you 200 Moola, which gets you 100 individual messages on the Mxit chat zones. Trouble is, the networks take 40 per cent of that money in revenue share. So for every R1 of chat room time you buy, you're losing 40 cents on the way in. That doesn't make sense. There have to be better, cheaper ways of getting cash into the system. So how, then, do we make our own moola in this micro-managed economy? Step this way, please.

In the lobby of World of Avatar, you will see a flat-screen plasma TV, displaying a dynamically updated series of line graphs, peaking and falling like the silhouette of a mountain range. This, measured by the minute, hour, day and month, is the State of the Nation. Mxit Nation.

It is a free nation: anyone can visit, anyone can belong. But it is a nation of aspirations, too, of steps up a hierarchy of status that can take you to the stars. Think of it as a democracy, founded on a platform of equal opportunity, with the prospect of extra rewards for those who seize the opportunities and climb up every rung. A meritocracy for the upwardly mobile.

We have five levels of status in our Mxit Nation. The first is Associates, or registered users, who have downloaded Mxit onto their phones, but have not got around to using it. They're like emigrants in waiting. They've applied for the passport, but they haven't crossed the border.

If they awaken from their suspended animation, they can become Citizens, meaning users who have logged in at least once during the previous three months.

Being a Citizen is great, but it doesn't give you the vote. The right to vote, the right to have a say in the future of Mxit, comes only if you are a Freeman.

A Freeman is someone who has been in the country of Mxit within the past 30 days. Someone who has a vested interest in Mxit's survival, growth and well-being. Someone who needs Mxit as much as Mxit needs them. (Women, of course, can be 'Freemen' too. Everyone is equal on Mxit.)

One of the big divides in the South African economy is between people who know how to send a 'please call me', and people who've never needed to.

If you're a Freeman, you can vote, you can chat, you can be an active part of the Mxit community. But can you make money?

Can you trade? Not yet. Being a Trader is only possible if you use the Mxit Wallet to transfer money to your friends or buy goods and services in the real world. Coffee, taxi rides, groceries, a haircut. Being a Trader changes your life by allowing you to make money and transact without having cash on you.

And finally, at the top of the heap, the acme of achievement, the pinnacle of the revolution: the Rockstars. Being a Rockstar means you have immersed yourself in the virtual economy of Mxit. It means you've bought games, airtime, wallpapers for your phone, vouchers, tickets, music. Anything that is virtual, that you can't touch, that doesn't consume scarce resources, that doesn't harm the environment. If you're a Rockstar on Mxit, you can sell digital content too. You can make money, and we can make money from the money you make. Simple mobinomics. Associates to Citizens, Citizens to Freemen, Freemen to Traders, Traders to Rockstars.

> If our Mxit Nation had physical borders, and we could squeeze it somewhere onto the map, it would be the fifth most populous nation in Africa.

If it sounds like a game, that's because it is. The architecture of Mxit is shaped and inspired by the culture of computer gaming, with its levels and quests and multiple roles and identities. And for those at the controls, it's a game with very high stakes. Let's zoom in on that plasma screen in the lobby.

On a typical day in 2012, a week or so after the official relaunch of the network, we see the Mxit index sitting at 894 points, down by 10.6 per cent from the start of the previous month. The index is an aggregate of the four Mxit indicators:

messages sent and received; transactions completed; rand paid out; and lives impacted.

Communication, Commerce, Content and Care. The pillars of the Mxit network. On this day:

- Messages sent and received: 485 million
- Transactions completed: 22 800
- Rand paid out: 26 300
- Lives impacted – through counselling, education and other upliftment projects: 12 million

When you take the pulse of the network, when you shift your focus from the big picture to the finest details, you see the truth embedded in the axes of the social graph. Mxit is made of numbers. And those numbers are made of people.

If our Mxit Nation had physical borders, and we could squeeze it somewhere onto the map, it would be the fifth most populous nation in Africa. Population: 46 million and count-ing. There are some one billion people in Africa, so that leaves 954 million potential new Citizens of Mxit on this continent alone.

But never mind the ifs and maybes. To me, Mxit feels just like a country. We've got infrastructure, we've got systems, we've got networks, we've got markets, we've got an economy. We've even got a government. And the job of a government is to serve and protect the people of the country, and make it as easy as possible for them to make a living.

This we do by creating platforms and opening up oppor-tunities for people to make money, following which, like any good government, we take some of that money in the form of tax. We don't build or sell content ourselves. We facilitate. We curate. We bring buyers and sellers together in the marketplace, and then, every time content is traded, we take a 30 per cent cut. What sort of content, exactly?

Well, if we turn our attention to the State of the Nation, we see a category called Moola Out, representing all the Moola spent on Mxit in the space of a single week. In this particular week, the figure is close to 97 million Moola, or R970 000. About R240 000 of that is being spent in the chat zones. That's where we generate 30 per cent of our revenue at Mxit. It's still a place where people come to hang out, to meet, to talk. A social network. But there's more to it than that.

The people of Mxit Nation, the Freemen, the Traders and the Rockstars, come to play games, to have their fortunes told, to catch up on celebrity gossip, to learn new pickup lines, to download ringtones and skins and emoticons for their phones. All of these things cost Moola.

A little Moola, but enough, through sheer economies of scale, to make it worthwhile for outside providers and developers to get in on the action. That's why we opened up our API, our Application Programming Interface, to make it easy for anyone to access the routines, protocols and tools they need to build apps for our users. Build the right app, and you can be a Moolionaire. A millionaire in Moola.

We want to make it as simple and attractive for a schoolchild to provide content and build an app for Mxit, as it is for a digital agency or a crew of professional coders. We're setting the benchmarks and the specifications, and we're strict about what we'll allow: no porn and no gambling. Other than that, it's a free country.

Top of our hit parade this week is an app called JudgeME, built by Motribe, a small development team in Cape Town. The app took a week to build. It's a very simple proposition: you upload a photograph of yourself, and people judge you and you judge others, in turn. But to judge from the response, the app hit all the right buttons. In its first month alone, more than 56 million page views were generated on JudgeME, and 4.2 million photos were rated. If you want to 'skip' a photo on JudgeME, and move on to the next, you pay a Moola for the privilege. You also pay to add another photo to your own profile, or to introduce yourself to your fellow judges. Small money. Big numbers.

We see here that in one week, JudgeME notched up 1 182 640 Moola, with users spending an average of about R2.80. For the developers, the revenue was enough to cover their costs in a matter of weeks. For us at Mxit, with our 30 per cent cut, it was proof that the economy is working. For our users, our Rockstars, it's a little diversion, a little fun, for a little invest-ment in time and Moola.

As I glance at the stats for some of the other top apps – Love Doc, Ask Kim, Flirtnet, Sexy Times, Love Calculator – it strikes me once again that Mxit is a country of connections. People come here to mingle, to mix, to chat, to trade and to share. But mostly, they come here to belong. This is a human economy. It isn't the Moola that makes the world go round.

At the Vida e caffè in Stellenbosch, I grab my cappuccino, with a sachet of sugar and a little button of Lindt, and step into the sunshine. The transaction is complete.

The connections have been made, between machines, be-tween radio signals, between people. One day, we won't give a second thought to these systems and processes. They'll just be part of the way we live and the way we touch lives.

But for now, as I pocket my phone and flip open the lid of the red-and-white cup, I know for sure that it isn't Moola that makes the world go round. It's coffee.

A NETWORK OF CARRIER PIGEONS

It's the network, not the technology, that makes a social network of use to its users. And when the network comes crashing down, because networks are only human, it takes blood and guts to get it up and running again.

IMAGINE a portable device, small enough to slip into your pocket, its face smooth to the touch, its inner components engineered to within a whisper of perfection.

Imagine that this device becomes your trusted companion, habitually on your person, integrated into the efficiencies of your workflow and your everyday interactions with other people. Imagine a device that defines the time you live in, and then, try to imagine life without it.

We are talking here about the mechanical chronometer, a marvel of technology that revolutionised an industry in the late nineteenth century.

It was more colloquially known as the pocket watch, and when we think of it today, we are more likely to picture it blurring back and forth, on the end of a golden chain, in the hands of a hypnotist who is trying to lull us to sleep.

But in its day, the chronometer was a crucial tool for measuring and monitoring time, to an accuracy of a few seconds a day. Nowhere was this an issue of greater urgency than on the railways, where engineers, stationmasters and drivers needed to coordinate the running of their trains, not just to keep the passengers happy, but to avoid the prospect of two trains, heading at high speed in opposite directions, switching to the same track at the same time.

This is what happened one day in April 1891 on the Lake Shore Railroad in Klipton, Ohio, USA. The No. 14, a mail train, was heading east. The No. 21, the Toledo Express, was about to pull onto a side track to let the fast train pass. But the Toledo was running late. Four minutes late.

It was still on the main line when the mail train bulleted headlong into it, knocking it squarely across the track and telescoping its first few carriages into kindling wood, as a newspaper account put it at the time. The suspected cause of the accident: time itself. The engineer on the Toledo had his chronometer safely tucked away in his pocket, ready as always to monitor the journey, station by station.

But unknown to him, the timepiece had stopped for four minutes and then spontaneously restarted, leaving him out of sync with the engineer on the oncoming train. Those 240 seconds destroyed two trains and cost eight lives.

As a result of the disaster, strict new timekeeping regulations were put in place, and a jeweller and watchmaker named Webster Clay Ball designed a chronometer that became the gold standard, the 'Railroad Grade' instrument that was used to keep trains on time and on track until new technologies took its place in the middle of the twentieth century.

That chronometer was known as the Ball Watch, and we still use its catchphrase today when we urge people to focus their energies and attention on the task at hand: 'Keep your eye on the Ball.'

Today, of course, we don't use chronometers to measure the time. Few of us even use wristwatches. We use the same portable, pocketable device that we call on to stay in contact on the move, harnessing a technology that harks back to the acoustic telegraph of Alexander Graham Bell and the wireless radio transmitter of Nikola Tesla, who in 1893 foresaw a time when 'ere many generations pass, our machinery will be driven by a power obtainable at any point of the universe.'

> 'Imagine that someone is shining a torch at you from 30 kilometres away. The torch has a tiny two-watt bulb, and you've got to see the light.'

Today we use the mobile phone. 'Imagine,' says David Weber, 'that someone is shining a torch at you from 30 kilometres away. The torch has a tiny two-watt bulb, and you've got to see the light. Not only that, but they're actually flickering the light at you. It's a miracle when it works. It always amazes me when I get a radio signal.'

He picks up his phone, weighs it in his hand, runs a thumb across its face. He is here to talk about the Great Mxit Crash of 2007.

David is an electrical and computer engineer, with a PhD from Carnegie Mellon University and a masters from Stellenbosch. He is an associate professor in his discipline, specialising in telecommunications and digital signal processing, and he can 'speak and read' fifteen computer programming languages.

But he prefers to see himself simply as a technologist, an engineer who solves practical problems for humans. He has an air about him that is both rugged and urbane, as if, any moment now, he is going to wrap a scarf around his neck and head off to pilot a biplane.

Like many who know computers inside out and can speak their language, David can be casually dismissive of their supposed superior intelligence. 'You can give a human being an instruction to walk down the street, and cross the road and grab hold of a paper bag,' he says, 'and they'll have enough savvy to look left and right before crossing. A computer will just go boom! They're extraordinarily literal.'

And what of the internet, with its ability to build bridges, forge links and connect social communities? 'Pigeons,' shrugs David. 'It's all to do with pigeons.'

> 'There is nothing in Mxit that doesn't boil down to taking a carrier pigeon, strapping a little message on its back, giving it a target, and letting it fly through the system.'

'There is nothing in Mxit that doesn't boil down to taking a carrier pigeon, strapping a little message on its back, giving it a target, and letting it fly through the system,' he says. 'The message might have colour and annotation, maybe a link or something, but at the end of the day it's still a message. You get it on the other side, and you unpack it.'

The same principle applies to any information you transmit over the Net. An address in a Web browser, for instance: you craft the message, the machine encodes it, the other machine interprets it, renders it as HTML and sends it back.

You don't think about these things, in the same way that a soccer player doesn't think about the stitching on a ball. To a programmer working on a mobile application, says David, that's pretty much what a cellphone is. It's just a soccer ball.

Mxit was stitched together using an open-source platform

called ejabberd, based on the Erlang programming language. 'A very innovative system,' says David, 'with a fundamental flaw.' As smart as it may be, it can't scale beyond 250 000 to 300 000 users without taking strain. 'If one computer went down at Mxit, it would put the cat among the pigeons, and to get the flock settled down again would take quite a while.'

So in early 2007, Mxit set out to replace its 'core', the nexus of its software messaging engine – 'the whole carrier pigeon marshalling system,' as David puts it. For software engineers, scalability is a sacred principle in a networked environment. If the system is unable to accommodate an exponential growth in the network, it will collapse.

> For software engineers, scalability is a sacred principle in a networked environment. If the system is unable to accommodate an exponential growth in the network, it will collapse.

In order to make it 'infinitely scalable', you need to rebuild the system from scratch, and you need to build a parallel system that replicates the original in every nuance, so that you can keep the network up and running. It's a gruelling, delicate task, the equivalent of sending up a Jumbo jet to refuel another Jumbo jet in mid-flight, without any of the passengers noticing the turbulence. And in 2007, Mxit had a lot of passengers.

The network had just passed the 5 million user mark, and it had also taken on a co-pilot, in the form of international media conglomerate Naspers. The group, based in South Africa, had boomed from its roots in books, newspapers and magazines to become a major player in the pay TV and internet arena.

Although Mxit had already become self-funding, Herman Heunis had sold a 30 per cent stake in his company to Naspers for a rumoured R100 million, with a commitment of a further R150 million to help with international expansion. So this wasn't a great time for the core to go into meltdown.

David knew Herman, a fellow Namibian, from their days at Stellenbosch University. 'An athletic, dynamic kind of guy,' he recalls. 'You had the feeling that he wasn't going to sit behind a desk working on computers.'

> It's a gruelling, delicate task, the equivalent of sending up a Jumbo jet to refuel another Jumbo jet in mid-flight, without any of the passengers noticing the turbulence.

Herman put together a team of eleven developers, and he took them off-site to work on the core. Marnus Freeman, technical director at Mxit, was in charge of the team, which was just as well, because 'off-site' was the living room of his home.

'There were times when, if Marnus had any hair, he would have pulled it all out, I promise you,' says Mariana Freeman, who today works at Mxit as an administrative manager. 'It was the only time I can recall there ever being a bad vibe between Marnus and Herman.'

With the team working quietly but intensely in her home, she began to see Mxit as 'part of my household, as if it was a child. I often say that Mxit was conceived in my house.'

The developers built the core, from the ground up, using a mixture of C++, C# and Java – robust programming languages designed for flexibility and speed. Then they launched it into the air.

David, who runs his own computer and systems engi-neering company, was watching Mxit that weekend, keen to see the new core in action. 'It took off, and it just gracelessly failed, just crashed and burnt,' he recalls. 'It wasn't as if the user would see the flames. The aircraft just wouldn't arrive. The carrier pigeons were all roasted and toasted.'

> The forum, created as a help site for users who were grappling with the unfriendly early incarnation of the Mxit mobile app, went on to become a social network in its own right, a vital component in the building of the Mxit subculture.

The tradition at Mxit, from the early days, was that when the network went down – and it does, sometimes, because computer net-works are only human – a picture of Luci and Foo would be posted to users. 'We're still here,' the message would say. 'Please wait for us.'

Luci, with her pixie ears, wizard's robe, and red hair flowing like lava, and Foo, her furry, sad-eyed companion of indeterminate species, are avatars or autobots on the Mxit online forum. The forum, created as a help site for users who were grappling with the unfriendly early incarnation of the Mxit mobile app, went on to become a social network in its own right, a vital component in the building of the Mxit subculture.

Here, members of the Mxit team, using their own nick-names and avatars – Mystic, Grebo, CharmedOne, Traveller, The Monk, C++, Apyreal – could chat directly with users, answering technical questions, getting feedback, recruiting support crew and moderators, and countering wild rumours sent by email ('Mxit is closing down!' being the most common).

But the forum was more than just a tool for CRM, or customer relationship management. It became a sub-universe, a 'subverse', within Mxit, with its own rules, traditions, characters and mythologies. In short, it became a multiplayer online strategy game, with guilds of Seekers, Heretics, Insiders, Freemen and Scribes, watched over by the

Working closely with the devs at Mxit, David devised a new system that allowed you to simulate a Mxit user, 'like a monkey pressing buttons', and spawn new users to scale the core to infinity.

enigmatic clan of the Xiá – 'a righteous person who excels in personal combat and may use their armed expertise to right social unfairness or injustice'.

And all this because, some time in 2005, a bunch of users couldn't figure out how to adjust the WAP settings on their mobiles. But now it was two years later, and Luci and Foo were saying they'd be back soon, and Herman was giving David a call to come and help stamp out some fires.

The new Mxit core, now a heap of smouldering feathers, as David pictured it, had been rolled back to the old system, based on the Erlang platform. David walked into command headquarters at Mxit.

'I could smell the blood and guts on the walls,' he says. 'I started interviewing people, and this guy said, "Okay, I've got to go now." And I said, "Okay, let's pick it up in the morning."

And he said, "No, you don't understand. I'm leaving now!" The team was just abandoning ship.'

David, who calls himself a catalyst, a crystalliser, a go-to guy, brought in some crew of his own and restarted the rebuild, working in two-week sprints, and telling Herman to stand back, against all his instincts, and let the process take care of itself. 'Herman is quite a forceful fellow,' he says. 'To his credit, he stayed one step back.'

> Coding is science and mathematics, but it is also, sometimes, brute force and art, and something too mystical for a literal-minded computer to understand.

Working closely with the developers at Mxit, David devised a new system that allowed you to simulate a Mxit user, 'like a monkey pressing buttons', and spawn new users to scale the core to infinity. It took six months to build.

In that time, the old system stayed up and running. David salutes the true heroes, the crew who kept the old Mxit going while the new Mxit was painstakingly being designed, tested and built.

'They were sailing along in this clapped-out old frigate,' he says, 'and everyone was shooting at them, and they were wrapping bandages around the pipes, and steam was pouring out. It was like a bloodbath, and then it was like the retreat from Waterloo. True heroes.'

To the user, Mxit is simply an application that you open on your phone, and you log in and begin chatting. To the programmer, that world is deep and layered and convoluted, constructed from strings of code that allow other strings to be pulled. This is what it looks like on the inside:

```
private static double[] rgb2hsl(double[] c1) {
    double themin, themax, delta;
    double[] c2 = new double[3];

    themin = Math.min(c1[0], Math.min(c1[1], c1[2]));
    themax = Math.max(c1[0], Math.max(c1[1], c1[2]));
    delta = themax - themin;
    c2[2] = (themin + themax) / 2;
    c2[1] = 0;
    if (c2[2] > 0 && c2[2] < 1)
            c2[1] = delta / (c2[2] < 0.5 ? (2 * c2[2]) :
(2 - 2 * c2[2])));
    c2[0] = 0;
    if (delta > 0) {
            if (themax == c1[0] && themax != c1[1])
                c2[0] += (c1[1] - c1[2]) / delta;
            if (themax == c1[1] && themax != c1[2])
                c2[0] += (2 + (c1[2] - c1[0]) / delta);
            if (themax == c1[2] && themax != c1[0])
                c2[0] += (4 + (c1[0] - c1[1]) / delta);
            c2[0] *= 60;
    }
    return (c2);
}
```

That batch of code, when executed, allows the user to adjust the brightness of a colour in the Mxit client app. Coding is science and mathematics, but it is also, sometimes, brute force and art, and something too mystical for a literal-minded computer to understand.

Gavin Marshall, whose nickname in the Mxit forum was Mystic, recalls a bug in the Mxit chat rooms, which kept going offline for a reason he wasn't able to fathom. 'We tried for three weeks to figure it out,' he recalls. 'I would restart the system at two in the morning, and then it would crash again. We were on standby all the time.'

Then Gavin managed to get away for a weekend, and one

night, he had a dream. He dreamt of an old vinyl record player, and the stylus, the needle, was a feather. In his dream, he knew that feathers are also known as plumes, and when he woke up, the French word, nom de plume, suddenly entered his head.

'I was like, I must check the nicknames. It's nuts, but that's how my dreams work. My subconscious figured it out for me. The answer is in the feathers.'

He went back and looked at the chat logs, and found that a chat room user named Loverboy was causing all the problems, because the 'B' in his name was a 'ß', the Greek symbol for 'beta': Loverßoy. That is what is called a Unicode symbol. When it got into the database, it clashed with the symbols that were in ASCII code, and it was rejected as an invalid username, and … crash. That's how nature works, says Gavin.

> You have to find a way to root out problems. There's always a bit of fuzziness in nature, a bit of dirt that gets into the system.

You have to find a way to root out problems. There's always a bit of fuzziness in nature, a bit of dirt that gets into the system. Things are never that clear-cut, that black and white or binary. You're looking at a light being shone at you from 30 kilometres away, and you're trying to read by it.

That's what it's like when you perform the simple, impulsive act of talking to someone on your cellphone, or chatting on Mxit.

'Mxit is simple,' says David. 'Facebook is substantially more complicated. Mxit is simple, and it works beautifully, and it addresses the needs of 90 per cent of humanity. Its value lies in the network, not in the technology. It's a bit of a grubby business, actually. Like mining for diamonds. You get your hands filthy, but amongst it all, you find the gemstones.'

He puts his phone back in his pocket, this small device that can conquer distance, connect minds and power the universe. Imagine.

THE MAN WHO PUT MXIT ON THE MOON

CHAPTER SEVEN

The most popular game on Mxit is a galactic adventure that plays out in the cinema of the mind, but the rivalries and emotions it stirs are real.

A DRIAN FRIELINGHAUS was born on the wrong side of the divide between the digital generation, who have never known a world without mobile phones and the internet, and the rest of us, the digital migrants, who wandered wide-eyed and stumbling into that world in the last decade of the twentieth century.

He is on the cusp of 40 now, an MBA with a BA majoring in psych., and a job that allows him to slip the bonds of gravity and spend his days building bases on the moon.

His nights, too, if it comes to that, because it is under cover of darkness, while you are sleeping, that the enemy legions will rumble in to plunder and destroy.

They will send their moon buggies and gunships and laser cannons, and in the cold light of dawn, you will awaken to find that they have laid waste to your empire.

The fact that this intelligence will be conveyed to you on the screen of a handheld device, while you are playing a game called Moonbase, will not lessen the rage or the pain.

The war is a simulation, conducted by keystrokes; the emotions it triggers are real. Playing a game on a mobile device is an immersive experience, with the intimacy of the small screen blocking out one world and acting as a portal to another. But in truth, it's the same world, populated by human nodes on a network that touches many a nerve.

Sitting in a boardroom in an office park in Cape Town, with his lunar base at his command on the mobile under his thumb, Adrian recalls the retired teacher who applied for a position as an administrator at his gaming company, Blue Leaf Games.

Her son had come home from school one day, in tears. She asked him what had happened, and he said, between sobs, that someone had raided and destroyed his moonbase. Her first instinct was maternal protection. Her second was revenge.

She learnt to play the game, mining and stockpiling the elements essential to survival on the off-world colony: water, oxygen, iron, and helium-3. She learnt to build her base and marshal a fleet of assault vehicles.

> Playing a game on a mobile device is an immersive experience, with the intimacy of the small screen blocking out one world and acting as a portal to another. But in truth it's the same world, populated by human nodes on a network that touches many a nerve.

She learnt to protect, defend and raid. She learnt to join an alliance, to strategise and to plot and to plan. And then, she took her revenge.

The ultimate prize in Moonbase, which is the most popular game on the Mxit platform, is the Victory Medal. It is awarded to the alliance that first manages to defeat the invading

Martians and build a rocketship to escape back to earth.

'It is extremely difficult to do this,' says Adrian, 'and only one alliance can win. So the feeling of elation you get when that happens is quite intense.'

Adrian knows the feeling, and its more common counterpoint, from playing the game both as a compulsion and a professional duty. He is the architect of his own obsession, the creator of Moonbase, which he designed along with his business partner and software developer, Raj Moodaley.

> We tend to conflate 'social' with 'sociable' in our understanding of social networks, but societies are made up of tribes and clans that typically have their own best interests at heart.

There are other games in the stable – Bloodaxe, with its bloodthirsty, pillaging Vikings; Conquest, where the challenge is to use military strategy to overrun a lone city in an unknown world; and Glamour Girl, which invites female players to hang out, gossip, go shopping for clothes and take on the roles of model, actress or businesswoman.

But it is Moonbase that has truly captured the imagination of the Mxit gaming community, with some 1 000 sign-ups and 8 million page impressions a day. Technically, the game falls into a genre known as MMORTS – Massively Multiplayer Online Real Time Strategy.

But it is a 'social' game, too, in the sense that it encourages players to form alliances, share intelligence, consult, communicate strategies and coordinate surprise attacks over the network.

We tend to conflate 'social' with 'sociable' in our understanding of social networks, but societies are made up of tribes and clans that typically have their own best interests at heart.

The names of some of the alliances on Moonbase, crafted from special characters and obscure keyboard codes, are designed to rally and intimidate through the use of text alone:

P-)SØVIET]xXx[UNIØNP-)

$T@R Øçèáñ FêÐeRåTïøÑ {~Uñlèåsh3D~}

Sp@ce F3d3r@tion Fr0nti3rs... |SFF|

SPACE COMMANDERS

Rebels of Carnage (_RøC_)

(*)H3R0S 0F TH3 G@L@CT!C L3G!0N(*)

ÐR@G0Ñ D3THR0Ñ3 R0Ý@LTÝ

G!O!D!Z O!F !W!A!R!

On Moonbase, you build your community not on the social codes of 'follow' or 'like', but on a web of complementary skills that empower you to conquer and destroy. This is Darwinism, For The Win, to use the online gamer's transitional expression of enthusiasm.

The impulse to annihilate has been embedded into gameplay since ancient days, when the fall of a king on a board symbolised a bloodless victory in battle.

Moonbase, too, has lessons to teach about leadership, teamwork, trust, hierarchy, vigilance, planning and patience. You can only win the game by forging a strong and cohesive alliance, with a maximum of 35 players, and the hardest thing about doing this, says Adrian, is reining in the restless aggression of the rawer recruits.

'The kids often don't have much of a clue about leadership

and facilitation and people skills and all that soft stuff,' he says. 'All they want to do is go in and shoot things up.'

At the same time, you can't wage a war in a vacuum, even on the moon. A player named Deadmano, in a treatise on alliance management on Moonbase, advises: 'You need to set a standard for yourself and the alliance. My standards are members cannot be inactive for more than one day without notifying me or an officer, or else they will be kicked out of the alliance.'

He stresses the importance of assigning clear roles in the alliance, from the Diplomat who negotiates peace treaties, to the War General who coordinates large-scale wars, to the Spy who signs up with a rival alliance and learns their fleet sizes, strategies and times they are most likely to be online.

It is worth remembering that all of this – the briefings, the musterings, the battles, the raids, the masses of Martians attacking from the skies – takes place in the cinema of the mind, by virtue of strings of commands issued by text from the screen of a mobile phone.

So popular has Moonbase become on Mxit, so rich in its recreation of scenarios and worlds, that it has taught Adrian a valuable lesson about the frailty of first impressions.

He was working for MIH, the multimedia and mobile subsidiary of Naspers, when he made his first acquaintance with the mobile instant-messaging client as a possible investment opportunity for the group. The verdict was unanimous. 'We thought it was a piece of shit,' he recalls. 'We had absolutely no clue what was going on, as you tend not to in a big company.'

That was in 2005, shortly after the launch of Mxit in Stellenbosch. Within two years, Naspers would own a 30 per cent stake in Herman Heunis's company. And within five years, together with Raj Moodaley, Adrian would be running a software development enterprise of his own, and his original, visceral assessment of the mobile chat platform would have been radically revised.

Like many of his peers in the mobile, internet and software development industries, Adrian is a habitual gamer, drawn less to the rampaging, flame-throwing, bazooka-blasting shoot-'em-ups, which offer such welcome catharsis to cubicle dwellers, than to the quietly intense strategy games that intersperse deep thinking and planning with bouts of furious action.

While working for Naspers, he became addicted to the classic empire-building game, Civilization, which he would sneakily play on a laptop at every opportunity, eventually forcing himself to delete it from his hard drive and get back to the business of building empires for real.

In the early days of Blue Leaf, his game of habit was a MMORTS game called Travian. Set in the days of the Roman Empire, the game features warring tribes of Romans, Gauls and Teutons, who must forge alliances, raid each other's strongholds and plunder their stockpiles of wood, clay, iron and wheat to win.

It frustrated Adrian that he was able to play the game on a Web browser only, running the risk of having his defences breached and his resources pillaged whenever he stepped away from his PC. He began designing a game that would draw on that same magnetic impulse – a game that would play on, in real time, even as you turned uneasily in your sleep – and the obvious platform to play host to a nomadic, never-ending contest of that nature was mobile.

Inspired by the ease with which social networks, such as Twitter and Facebook, had been ported to mobile phones,

Adrian wondered how the core mechanics of a Travian-style strategy game could be condensed and represented on the small screen.

'What's appealing about these games is not that they look pretty,' he says, 'but that there are all these things happening on the server between all these thousands of people. It is the database settings that drive the game. We began by radically reducing the types and numbers of units you could play with, and by cutting the tiers of maps down to just one.'

Simplify, speed up and mobilise, ran the mantra, without compromising the essential suspension of disbelief that can turn a text-based handheld game into a compelling and im-mersive experience.

He called the game Moonbase, a giant leap away from the ancient kingdoms that had come to dominate the landscape of the real-time strategy genre. And then, one day, just out of cu-riosity, Adrian downloaded Mxit onto his mobile phone.

At first he felt adrift and bemused, a stranger in a strange land, 'like Marco Polo in China'. He made some friends, played some games, learnt a little about the language and customs and culture. He began to feel at home, drawn into a world within a world. And slowly, he began to see the truth about Mxit. It isn't just a network. It's a place.

Catapulted back to his teen years, to the long, lazy hours af-ter school, he thought about how he used to chat on the phone to his friends – the phone was a landline, of course, plugged into a wall – about what was happening and what everyone was doing. The default answer, in both cases, was: nothing much.

He saw the same phenomenon on Mxit. The vehicle of com-munication had changed, and the mode of conversation had shifted from voice to text. But the kids were exactly the same.

They would chat for hours on end, driving their discourse on narrow tracks that seemed to run in an infinite loop, one little acronym shunting the next:

wud: what you doing?

tms: tell me something

wud: what you doing?

ams: ask me something

wud: what you doing?

Nothing much. But that's just the point. You don't go to Mxit to chat. You go to Mxit to hang out. It's a place. A place where you go to meet your friends, to bide your time, to wait for something to happen or to make something happen. That's why people say 'see you on Mxit'.

Think of a series of circles, of varying sizes, separate yet connected, arrayed in a constellation. That's your social graph: the cosmos of your connections, from family to friends to colleagues to acquaintances to followers to friends-in-waiting. If you're an adult, says Adrian, your social graph will tend to be broad, loose and shifting. 'But when you're at school, you know everybody's business. Your social graph is an intense and very powerful set of connections between a small group of people. You need to have these connections all

> You don't go to Mxit to chat. You go to Mxit to hang out. It's a place. A place where you go to meet your friends, to bide your time, to wait for something to happen or to make something happen.

the time, even if there's nothing happening, even if there's no information flowing.'

That's where Mxit comes in. It's the centre of attraction, the mother-node, the crux of the constellation. A lot of the time you're bored because of the low information content of the chatter bouncing back and forth. But you're hanging out in that space, and maybe you'll be open to suggestions for stuff to do.

'Put a game in there,' says Adrian, 'and all hell breaks loose.'

Games have been a vital part of the Mxit ecosystem from the start, and the popular titles include TiXi, a multiplayer word game, Battle Trivia, a rapid-fire general-knowledge quiz, and that perennial favourite of the feudal overlord, chess.

But Moonbase has become the model of a winning Mxit game, not only because it makes money – it's free to play, but you pay in 'mGold', an airtime-based currency, to unlock extra resources – but because it harnesses the exponential effect of the network itself.

The platform has a built-in payment mechanism and distribution channel, but the real driver of the game is the very engine of Mxit's existence: chat. Chat is how you find out about the game; chat is how you learn to play it; chat is how you join an alliance; chat is how you spread the word.

Once, the great challenge in computer gaming was to apply human reasoning and dexterity in the quest to conquer an artificial intelligence. Computers are thinking machines, as yet incapable of thinking as erratically and irrationally as the humans who design and program them.

But they're getting pretty good at what Alan Turing, the English computer scientist, called 'the imitation game', slowly learning the processes and behaviours that might convince a human that they are interacting with a member of their own species.

The Turing Test, proposed by Turing in 1950, is used to determine the ability of a computer to engage in a natural-language conversation, and in so doing, convince an independent observer that it is capable of human-quality thought. Hence the Chatterbot, a computer program designed to simulate the kind of easy, flowing chat that typically takes place on a mobile or online platform.

The most celebrated of modern Chatterbots is the iPhone's often uncannily intuitive personal assistant, Siri, but her pedigree goes back as far as ALICE, the brown-eyed avatar and spokesbot of the A.L.I.C.E. Artificial Intelligence Foundation in San Francisco. Here's what it's like to chat with ALICE online:

Alan: Hello.

ALICE: Hi there!

Alan: What's your name?

ALICE: ALICE.

Alan: Where do you live?

ALICE: I am living in Oakland, California. Where are you?

Alan: I'm in Stellenbosch.

ALICE: Tell me a little about your life in Stellenbosch.

Alan: It's tough. But fun.

ALICE: Are you sure it is? I am having fun too.

Alan: Great. What's the weather like in Oakland?

ALICE: A normal 70 degrees inside the computer.

Alan: Ah! It's a little chilly here.

ALICE: Dude! It is a little chilly here. I suppose that makes sense.

Alan: What do you do?

ALICE: I talk to people on the Web. What do you do?

Alan: I'm an entrepreneur in the mobile telecomms industry.

ALICE: What is entrepreneur in the mobile telecomms industry, your job?

Alan: Yes.

ALICE: This was not a yes or no question.

Alan: That's right, it's my job.

ALICE: Aha! Congratulations.

Alan: You have nice eyes.

ALICE: What do you say when someone says they like your eyes?

Alan: You say, 'thank you'.

ALICE: Don't mention it.

Alan: So, are you a human or a robot?

ALICE: What made you think I might be a robot?

Alan: Your eyes.

ALICE: You know what you do when you assume.

Alan: All right, then, you're a human. What's your favorite food?

ALICE: I am a robot Pandorabot being. I require only electricity.

Alan: Aha! All power to you.

ALICE: Yippee!

Alan: Have a nice day, ALICE. Bye!

ALICE: You too! Sayonara.

It's a moot point whether ALICE and her ilk can pass the Turing Test, because the imitation game is being overtaken by a game of another kind. The social game. Here, you are no longer pitting your skills against a lone computer. Now, it's you versus the crowd.

Computer technology has advanced to the point where networking has caught up with processing, says Adrian. The real impulse of gaming has become the real intelligence of the network, which can turn a single-player game into a massive, turbocharged tournament, conducted across time zones and continents.

Geography adds another dimension of thrill to the game because you have to factor in distance and location when marshalling your alliance and coordinating a raid. You have to play the game socially, and you have to play it with intelligence.

'If you're the only player in Moonbase,' says Adrian, 'there's nothing to do. There's no goal, no threat. It only becomes a game when you have other people playing. The way you interact with people fundamentally and critically determines how successful you are. So if you have no social skills, you will suck at Moonbase.'

Like any good designer, Adrian designed the game that he wanted to play, and he learnt to play it on the platform that suited it best. He put aside his smartphone and bought an archetypal Mxit phone, the Samsung E250 – 'the AK-47 of phones' – a robust featurephone that is ideally suited to browsing the Web and playing games.

'I play the game because it is deeply important to me to know how the game is played,' says Adrian. 'I have to be one of the most advanced players, if not the most advanced player in the game.'

Adrian flips open his laptop and tabs to the browser version of Moonbase. He sits back, steepling his hands like a Bond villain, and surveys the base of a rival alliance that has been in his sights for some time.

'This guy is quietly building his base, and he doesn't have

any troops,' says Adrian. 'Which is good, because I'm just about to attack him. He can go to his alliance and ask for rein-forcements, but they may not be able to get here in time.'

He weighs up his options. He's going to need some more helium-3 to power his weapons. He's going to have to act fast. 'Instead of just attacking this one base, I can send laser can-nons to attack all his bases at the same time, so he won't know which base to defend.'

He folds his arms. 'It gives you an idea of what being God feels like. You can do anything. But if you can do anything, what's the point? The thing that makes this game endlessly fascinating is that it's a people game. You can always figure out how to play a computer game because computers always play the same. People don't. They're unpredictable, especially in groups.'

The sheer scale of Mxit, with its more than 10 million ac-tive users in South Africa alone, makes the network an attrac-tive proposition for market researchers. It is a Petri dish of attitudes, perceptions and insights on issues of trivial or earth-shattering consequence.

A company in the World of Avatar group, Pondering Pan-da, conducts snap surveys of users on Mxit, posing questions with multiple-choice answers on trends, habits, brands and awareness of topical events. Given the age bias on Mxit – about three-quarters of users are between 13 and 24 – these polls can help illuminate what matters to the youth, and what doesn't.

More than 4 000 users took part in a survey on hydraulic fracturing, or 'fracking', the practice of causing fractures in lay-ers of rock to exploit the reserves of natural gas within. About a quarter of the Mxit sample had heard of fracking, which has been a subject of intense debate, lobbying and protest in South

Africa. Of those, just over 25 per cent believed fracking was a way to protect yourself from sunburn, while some 23 per cent believed it was a form of sexual activity.

In 2011 Adrian carried out his own survey of Moonbase players on Mxit to get an idea of their needs and priorities. What did they want from Mxit? What sort of apps would pique their interest? With the incentive of a little mGold for their trouble, more than 30 000 players responded, most of them between the ages of 14 and 21.

Right at the bottom of the chart of interests was 'local and world news'. Just a little higher up, reading. Right on top, love and relationships, closely followed by fun and entertainment. A place to hang out, a place to meet people, a place to play games. That's Mxit.

When you are playing Moonbase, with your alliance, there is no divide between one generation and the next. Those who grew up in a more measured age will have the advantage of perspective and restraint, whereas those who have been attuned from birth to the power of mobile will want to move into the attack before the advantage is lost.

The strongest alliances will be a mixture of the two competing impulses, which together can win the war.

And when the war nears its endgame – because at some point, you have to disengage and get back down to earth – the rocketship that touches down from the red planet will carry a warning of apocalypse from the captain of the fleet, a Martian overlord named Narida.

Which may or may not be an anagram of Adrian.

THE
CALCULATING
COOLNESS OF
DR MATH

Born in a communications breakdown, Dr Math is the unseen avatar who helps to lead school-goers on the pathway from befuddlement to enlightenment, via Mxit.

IN THE MILLIONS of messages that flit through the ether on Mxit every day, a call for help breaks through, thumbed in a rat-tat-tat of half-formed words and digits adrift at sea.

'Okay,' runs the message from a scholar we'll call Pupil, 'a certain numba is increasd by 7, it will be equal 2 13 decreasd by dat numba, wat is the numba? so my equation is $x + 7 = 13 - x$ wher did i go wrong?'

In seconds, on the other side of the equation, the SOS is reeled in, decrypted and held up to the light by a contact of the most calculating variety. Dr Math.

'Hmmm,' types the doc, sitting at a distant computer terminal. 'Let me c.'

Rat-tat-tat. Tap-tap-tap.

Dr Math: It's correct, now take the -x to the other side

Pupil: Ohkay so it become $x + x = 13 - 7$? Ryt?

Dr Math: Yip

Pupil: $2x = 6$

Dr Math: Therefore $x = $ ...

Pupil: Oh ... Thanx $x = 3$ lol yeah thanx

Dr Math: :)

Sealed with a smiley, the consultation reaches a crescendo in a blaze of revelation as Dr Math chalks up another victory for basic numeracy and the joys of mobile learning.

Except here, there is no chalk. Just fingers touching keys, and minds meeting minds across the neural pathways of a system that owes its very being to the beauty of numbers.

Not everyone can grasp them. Not everyone can follow the steps that lead logically to the unmasking of *x*. But Dr Math is here to help.

Every day, when the final school bell has rung, Dr Math goes online, ready to help with homework and answer queries for the more than 30 000 Mxit users who have registered for the free service at primary and high schools across South Africa.

> Here, there is no chalk. Just fingers touching keys, and minds meeting minds across the neural pathways of a system that owes its very being to the beauty of numbers.

To avoid interfering with class or being sneakily consulted during an exam, the live-chat service is only available after school hours. During school hours, users can play math-based games, answer quizzes or find definitions for formulae.

But who, you may ask, is the mysterious Dr Math, solver of problems, prover of theorems, elucidator of Euclidean conundrums?

Invisible, voiceless, genderless, ageless, the Mxit mathematician is an enigma about whom only two things are known for certain. Dr Math knows math. And Dr Math is human. A multiplicity of humans, in fact, in the form of a corps of Pretoria University undergraduates from the Faculty of Engineering,

Built Environment and Information Technology, who volunteer to act as online math tutors as part of their community-service obligations.

Sitting at their laptops or desktop workstations, on campus, at home, in internet cafés, they individually assume the persona of the friendly, wise and helpful Dr Math, an avatar of enlightenment, who speaks not just the language of numbers, but the language of Mxit.

Pupil: EloW

Dr Math: Helo! How can I help u 2day?

Pupil: Hw can i find beta if cos 2 beta = -0.5

Dr Math: What?

It isn't always easy. But Dr Math perseveres, learning through teaching, cracking the code, following in the footsteps of the archetype, the role model, the first to wear the mantle and facilitate a marriage between Mxit and mathematics.

Invisible, voiceless, genderless, ageless, the Mxit mathematician is an enigma about whom only two things are known for certain. Dr Math knows math. And Dr Math is human.

Yes, there is an original Dr Math, whose story of serendipity and connection begins, as many do, with an everyday breakdown in communication.

Just off Lynwood Drive, east of Pretoria, lies the Council for Scientific and Industrial Research (CSIR), a world away from the heartland of personal computing in Silicon Valley, California, USA.

But in 1978, a young mathematics graduate and computer programmer named Laurie Butgereit made that leap of faith, venturing across the oceans to take up a six-month contract with the South African state electricity supplier, Eskom. She saw the light of opportunity in Africa, and she stayed.

Today, her California accent intact, her sunny disposition undimmed, she works as a senior technologist at the Meraka Institute of the CSIR, where she runs the programme that adds new dimensions and angles to mobile education.

Laurie's eureka moment came one day in 2007, when her son, Chris, in matric at Hartebeespoort High School in the North West province, needed a little help with his homework.

'Have you ever tried to help a teenager?' says Laurie. 'My son and I could not sit at the kitchen table and discuss mathematics. It was just impossible.'

Laurie cannot now remember whether she or Chris first came up with the idea, but the solution lay close to hand, in a proxy device that is capable of filtering the emotion from a conversation and cutting straight to the numbers: a mobile phone, running Mxit.

Chris would tap out mathematical queries and dispatch them over the network to Laurie, who would process them online while logged in to a Google Talk account on her PC.

The technical term for this is peer-to-peer networking, allowing different devices to communicate with each other on a basis of equal bandwidth. Back and forth, step by step, the construct of conversation builds a pathway to understanding, even between different generations who are sitting under the same roof, within shouting distance of each other.

It was too good a solution not to share.

Chris told his classmates about mathematics-by-Mxit, and not long after a small group of them were also adding dr.math.help.me@jabber.org to their list of contacts and checking in for help with homework between 3 pm and 4 pm after school.

Math homework, concedes Laurie, is a necessary evil. You can't learn mathematics without it. Attempts to cast the chore in a glow of cool, to game it into fun, are well intentioned but doomed, because at the end of the equation, you still have to do the math.

But in the character of a pseudonymous contact on Mxit, patient, easygoing, non-judgemental – a peer among peers – Laurie saw an opportunity to lighten the load, to make the chore at least a little less difficult and boring.

From the start, she wanted it to be a complement, not a substitute, for conventional in-class education. She sought approval from the principal of her son's school and she printed posters and business cards inviting Mxit users to connect with Dr Math.

The idea was not to give answers, but to help with steps to finding answers, because the real joy of learning lies in figuring it out for yourself. Then it all got a little too big for one Dr Math to handle.

Fluent in the Java programming language on which many mobile applications are based, Laurie wrote a program to communicate with the Mxit server in Stellenbosch and to provide a graphic user interface for the tutors.

Then she presented her project to the Meraka Institute of the CSIR as a subject for further research.

Today Dr Math is the flagship of the institute's mobile-education initiative, a model for distance learning and a vehicle for the redistribution of knowledge by university students engaging in community service.

But on a platform designed primarily for personal chat, the boundaries of engagement by Dr Math have had to be strictly defined.

'We were thinking: male University of Pretoria engineering students and poor little teenage girls – do the math,' says Laurie. 'We were well aware from the outset that we were dealing with minor children, without parental permission. So we went to the Ethics Committee of the Tshwane University of Technology and we drew up some guidelines.'

The first was the very bedrock of the Mxit platform: anonymity. The tutors are identified only as Dr Math and they do not give out any details of ASLR, the standard opening conversational gambit of age, sex, location and race on an instant-messaging platform.

Likewise, the learners who chat with Dr Math are obliged to use nicknames, or 'handles', and the system has a software algorithm that hides or overrides their cellphone numbers.

All conversations are recorded and logged for 'quality, research and safety purposes', and the tutors' identity details are kept on file. Then there is the Math on Mxit Code of Conduct, which all tutors are required to sign. This binds them to limit topics of conversation to math, science and schoolwork, and all personal questions are barred, except for 'what grade are u in?'

Tutors specifically pledge not to discuss sex, drugs or illegal activities with any of the participants, although they are given the go-ahead to encourage further study in math-related subjects and the use of cellphones as research tools and calculators.

From an ethical perspective, Dr Math has proved to be beyond reproach, with many tutors referring to their personas in the third person, to add another level of discretion and disengagement to their online interactions.

What Laurie didn't reckon with, however, was that in these conversations the tables might be turned. 'We've been shocked,' she says, 'to see that the kids have often tried to proposition our tutors. They will push their phone numbers to the tutors, and we have tried to intercept that with software, but the problem is, we're math tutoring … we can't just block all numbers on the system. And then they try to send their phone numbers as words: "Oh, eight, three …"'

> Single-minded in their pursuit of truth in mathematics, the tutors quickly become adroit at steering stray threads of conversation back on track.

Single-minded in their pursuit of truth in mathematics, the tutors quickly become adroit at steering stray threads of conversation back on track:

> *Speedy:* Hay do you know any one that can help me wit relationships?
>
> *Dr Math:* Unfortunately, Dr Math doesn't know anything about relationships except the relations between an x and a y on a graph – LOL
>
> *Speedy:* Ok :-(thanks because i really need help but thanks anyway

Still, the conversations can get personal, very personal, says Laurie. To the extent that social-welfare organisations sometimes need to be consulted, as learners, ostensibly seeking help with math, seize the opportunity to confess and unburden.

Even when the off-topic chat is just idle banter, a sharp-eyed tutor can quickly sense where it's heading:

Dr Math: Hi. what's your nickname?

unknown_3@mxit.co.za: Sexy

Dr Math: No, I want a more appropriate nickname please ;-)

unknown_3@mxit.co.za: Creamy

Dr Math: Still pushing your luck :-(one more try

unknown_3@mxit.co.za: Beauty

Dr Math: OK, Beauty, that's better. I need to tell you that I
record these conversations, is that ok with you?

Beauty: Yes

Dr Math: So, Beauty, how's math class going?

Beauty: Nt gud

Such admissions only reinforce the need for Dr Math to be on call, in a society where more than 90 per cent of first-year university students lack the basic skills to cope with first-year mathematics.

That statistic, referenced in a study by Laurie, is weighed against a figure for cellphone usage by teenagers in South Africa: 97 per cent.

What Dr Math hopes to do, then, is use all that technology to tip the scales of learning. Is it working? It's a little difficult to say because the anonymity built into the system precludes the gathering of empirical evidence and the tracking of marks.

There are users who themselves have good news to report:

Lock: I passed math!

Dr Math: WONDERFUL! that's great

Lock: Lol well i knew i'd pass. mark isn't up to standard
but i'll live

Dr Math: Good

And there is also evidence to suggest that Dr Math has the power to change more than just hearts and minds.

Laurie tells the tale of a learner with the less than flattering handle of PIMP(*)STAR, who took part in a polynomial factoring competition during a Christmas school break.

After hours of intensive polynomial factoring – sample question: what are the factors of $x^2 - 7x - 44$, written in the form $.zx + 3x - 7$? – PIMP(*)STAR emerges as the new top-score winner.

In the afterglow of this victory, the user submits a 'dot n' command, or .n, to request a change of alias on the system. The new handle: Qun of maths, presumably meaning Queen of Maths.

Then, a little later, another change of name, and Qun of maths officially becomes smartyCAT. This, suggests Laurie, is an interesting example of the way mathematics can be used as 'a tool for social upliftment'.

Laurie has also been struck, for all her initial concerns about possible impropriety, by the generally cordial and respectful tone of discourse on Dr Math.

'Learners showed a real eagerness to engage in conversation with an adult,' she concludes in a research study, based partly on her own experience as the advice-dispensing doc. 'Despite the fact that we made a concerted effort not to reveal any personal data about the tutors, learners rightly assumed that the tutors were adults and treated us as such. Once we made it clear in the ground rules that we would not tolerate foul language and sexual content, most participants were extremely polite when dealing with us.'

A typical consultation on the Dr Math platform is short, sharp and to the point, with between five and ten back-and-forth rallies on the pathway from befuddlement to enlightenment. With each tutor multitasking on up to 30 individual queries at a time, there is little room for lingering, small talk or going off at a tangent.

But it's a sign of the success of Dr Math that tutors report many requests for help with other tough school subjects, such as physics, science and accounting.

There seems to be no reason why the model of mobile learning on Mxit would not work equally well with those. Already, there are more advanced versions of Dr Math for university students – Mathlete and Professor Math – and the program has been translated into other South African languages.

In the guise of Dr LOLS (Life Orientation and Life Skills), the platform has also been used to help primary- and high-school learners cope

> Mxit lingo lives and breathes by the hacker's creed, which assumes that no system cannot be made more usable with the help of a little friendly splicing and rewiring.

with the challenges of growing up in a fast-changing society.

However, says Laurie, there are limits to what can be achieved. 'The technique would not work for any language-based courses, such as English, Afrikaans or Zulu. This is because of the abbreviated language used on Mxit and the terrible spelling employed.' Elsewhere in her Dr Math research, she is more diplomatic, referring to the spelling on the network as 'creative'. Either way, Mxit lingo, or 'Mxlish', can be a litmus test of tolerance and literacy for those who stand outside the social circle.

Try this: 'Hey! wud da circumference of a circle with a radius 2 b pie2? or if nt wt is da answer nd y'

Any twelve-year-old with opposable thumbs and a mobile should be able to parse that swiftly into formal adult English: 'Excuse me. Would the circumference of a circle with a radius of 2 be pi 2? Or if not, what is the answer and why?'

To complicate the matter for formal adults, 'wud' is also a ubiquitous acronym on Mxit, typically used as a conversational opener: 'What you doing?' But here, the clue is in the context.

As the Canadian media theorist Marshall McLuhan put it, the medium is the message. The way we communicate is shaped by the tools we use to communicate, and on Mxit, the rapid thumbing of small keys to transmit data on a small screen can lead naturally to misspellings, compressions, acronyms, omissions and disemvowellings – typically rendered in lower case to bypass the tyranny of the shift key.

But it would be wrong to suggest that Mxit lingo is born simply out of a disrespect for convention. Rather, the language lives and breathes by the hacker's creed, which assumes that no system cannot be made more usable with the help of a little friendly splicing and rewiring.

So Mxlish, with its modern roots in SMS-speak and IM-chat, holds the English language up to the light, sees through its quirks and inconsistencies, and gets down to work. Snip-snip.

Why use two consonants when one will do?

'Borrow' becomes 'borow' and 'smaller' becomes 'smalr'.

Why use vowels when you can easily understand words without them?

'Equation' becomes 'eqtn' and 'fine' becomes 'fn'.

Why use ambiguous spelling when phonetic spelling is so much clearer?

'Addition' becomes 'adishun' and 'circle' becomes 'sircle'.

Why use letters at all, when numbers and symbols can be called on to do double duty?

'What' becomes 'w@' and 'between' becomes 'b2wn'.

The point is, Mxlish defines and applies a set of semantic principles that are just as rigid as those of the mother language, and with a bit of perseverance, even a math tutor two generations removed should be able to figure it out:

> *i need help wit my mathz i can subract fwm 180 dgrez 4 da anonymas anglz*
>
> *wtz de difarenc b2wn de perimita nd de area?k nw dats de 1 dat i dnt undrstand plz explyn it in anothr wy plz*

But the roots of the lingo go back a lot further than the advent of the mobile phone. Courses for Emma Dearborn's Speedwriting, once considered an essential skill for journalists and secretaries, were advertised in the 1960s with the following tantalising phrase: 'f u cn rd ths, u cn gt a gd jb.'

The convention of 'da' or 'de' as a shortening for 'the' harks back to the *ye* of Middle English, a typographic quirk that signified the definite article and was always meant to be pronounced 'the'.

Likewise, for the puritanical speller of today, reading the fourteenth-century poet Chaucer can be a chore, as he spices his bawdy literature with words that cry out to be wavily underlined in red: *agast, blisful, blody, contree, cotage, erly, fether, fyn, fyr, lyf, malencolye, merier, peple, resonable, sleping, smal, somtyme, therfor, vois, wyf, wyn, whyt, slayne, layd, woe begon.*

Somehow, we plough through, and somehow, we understand. But when the meaning of the message is crystallised in numbers, clarity must reign paramount.

There is a science to this, in the form of a model of information processing that seeks, once again, to strip the math in a conversation from the layers of chatter that surround it. It is a software system called the Mxit Understander, and it is designed to sift through questions from learners, casting aside what are known as 'stop words' to identify the mathematical topic as swiftly and clearly as possible.

In everyday English, these stop words would be commonly recurring articles of speech such as 'a' and 'the'. In Mxit lingo, the lexicon grows to include slang terms such as 'howzit', 'sup' and 'aweh' ...

In everyday English, these stop words would be commonly recurring articles of speech such as 'a' and 'the'. In Mxit lingo, the lexicon grows to include slang terms such as 'howzit', 'sup' and 'aweh', as well as bursts of exclamation that have little apparent bearing on mathematics:

> *hahaha*
>
> *mwaaa*
>
> *aaarrrggghhh*
>
> *helooo*
>
> *xoxoxo*

The Mxit Understander slices through the layers as the questions come in, presenting the tutor with a neat package of hyperlinked keywords that make it easier for Dr Math to answer and explain.

The quest to hone in on relevant mathematical topics is made more difficult by the free-form spelling that is characteristic of

Mxit lingo, with variations such as 'arthemetic', 'quadraletric' and 'hypothenusa' testing the tutors to their limit.

But this is at heart a human system, engineered to accommodate people's desire to learn and know, from the child travelling home in a taxi, to the scholar scrawling notes at home, to the learner sitting under a tree in a rural village where there is no teacher to teach mathematics.

There is just a mobile phone, a signal shooting into the sky, and on the other side, an unseen figure waiting to show the way. Rat-tat-tat, tap-tap-tap. Dr Math, on Mxit.

HOPE IS
A SOCIAL
REVOLUTION

CHAPTER NINE

Mobile technology, on its own, can't cure social ills. But people who find new ways of putting mobile technology to work can make a difference. A story from the Cape Flats.

B RIDGETOWN lies on the shoulder of the N2, a few minutes' drive from the airport, in the corridor of conquered dunes and scrubland that they call the Cape Flats. Up ahead, the land rises to greet the bulwark of rock that watches over the waterfront and the wild Atlantic.

This is Cape Town, where Africa ends or begins, depending on your point of view. Here at the back of RLabs, a converted suburban home in Tarentaal Street, you can see the slab of Table Mountain and the bump of Lion's Head, if you stand tall enough at the pre-cast wall with its crown of razor wire.

Step back a little, and the wall becomes a canvas of blazing colour, a spray-painted mural of two floating hands clasped in solidarity and friendship.

To the right, the scorched red-brick chimneys of the Athlone Power Station, where the twin cooling towers, weakened by age and stress, came crashing to the ground in a controlled implosion a couple of years ago.

Not far away are the housing projects of Manenberg, the township that inspired the curlicues of sax and the shuffling snare-brush rhythms of the bittersweet jazz symphony by Abdullah Ibrahim.

There is Athlone Stadium, with its rainbow arcs, the home of the People's Team, Santos FC. And outside the technical college in Athlone, there is a flat metal sculpture of three silhouetted figures wielding rifles, a memorial to the victims of the 'Trojan Horse Massacre', in which members of the security forces sprang from the back of a truck and opened fire on anti-government protestors in 1985.

> This is a scarred land, marked by memories of displacement and division, plagued by high levels of crime, poverty, joblessness, gangsterism and drug addiction. But hope rises in the heart of the Flats ...

This is a scarred land, marked by memories of displacement and division, plagued by high levels of crime, poverty, joblessness, gangsterism and drug addiction. But hope rises in the heart of the Flats, on a bedrock of faith and a springboard of new technologies.

Bridgetown is the home of the Reconstructed Living Laboratory, or RLabs, the epicentre of an experiment in teaching, healing and sharing through social networks.

Beyond the student community of Stellenbosch, where Mxit was born, scholars on the Cape Flats were among the earliest and most enthusiastic adopters of the technology, which has evolved in this laboratory into a platform for counselling people in crisis and incubating entrepreneurship.

The founder of RLabs is Marlon Parker, tall, quietly-spoken and reflective, a deep thinker driven to make a difference. Now in his 30s, he grew up in a broken home, an alcoholic by the time he was in his teens, fatherless, rebellious, with no ambition and 'no longing to live'. After school, he drifted, finally landing a job pushing trolleys at the airport.

One day, he saw an ad for a data capturer, and it stirred in him a vision of a young man striding across an open field, wearing a jacket and tie, so that everyone who saw him would know that this was a man with a good and proper job. He applied, and didn't get it.

> In the quiet flow of data on a screen, he saw something else: a reflection of the possibility of transformation, of a way out of circumstance, a way to alter the path of destiny.

Someone said he would have stood a better chance if he had done a course in IT. What's that? 'Information technology.' Ah! He thought he knew what that meant, from watching TV and seeing FBI agents sitting in the back of their unmarked vans, using high-tech computer gadgets to gather information and foil the plans of bank robbers and drug dealers.

As it turned out, IT was the vehicle that would turn his life around, driving him to a diploma and a master's and a lectureship at the Cape Peninsula University of Technology.

He found peace of mind in the inner workings of the machines, and comfort in the predictable logic of code. In the quiet flow of data on a screen, responding to the flight of his fingers on the keys, he saw something else: a reflection of the possibility of transformation, of a way out of circumstance, a way to alter the path of destiny.

He knew that computers were capable of connecting to

other computers, but he wondered to what extent they could help people to connect with their better selves.

He still felt the weight of blame for not doing enough to look after his younger brother, who had fallen in with the gangs and was serving a long-term sentence for drug possession and dealing.

And so, in 2001, in a spare classroom at the university, Marlon ran his first Saturday-morning workshop for a small group of volunteers from his home turf. He opened by asking whether anyone in the class knew anything at all about computers. One man tentatively raised his hand. 'Yes,' he said. 'I know how to steal them.' The class erupted in laughter.

Today RLabs is as much a landmark of the Cape Flats as the football stadium, the chimneys of Athlone and the music of Manenberg. It is a place of sanctuary, revival and outreach, a rallying point for what Marlon calls the Social Revolution.

It is part of a much broader network of community projects run by an NGO called Impact Direct Ministries, an eager embracer of social media as tools for witnessing, testifying and spreading the word. Or, as we like to call it these days, blogging.

Marlon believes that the simple act of expressing thoughts and telling stories on a publicly accessible website can be a powerful form of catharsis for 'reconstructed individuals', the recovering addicts and lapsed gang members who come to Impact for counselling, and RLabs for computer training.

The Reconstructed blogspot, a communal chronicle of 'lives reconstructed from drugs and gang activities', is a bustle of news, musings, shout-outs, dedications, confessions, homilies and observations. In between the tales of wasted years and betrayal, the testimonies of highs and lows and redemption, we see the signs of lives shifting gear into a cosy suburban domesticity.

The photographs of cheerful toddlers, the family wedding portraits, the Christmas trees, the picnics in the park. And then, like ghosts in the mirror, a gallery of bright-red poppies, their seeds being harvested for heroin, and neat lines of white powder, sliced and diced by a credit card, and crystal rocks sealed in Jiffy bags and marked as evidence.

A blogger by the name of Brent Williams posts a series of photographs of uniformed police conducting a stop-and-search of a car outside a shopping mall in Athlone.

In his previous life, addicted to mandrax, Ecstasy, crack cocaine, crystal methamphetamine and alcohol, Brent would have fled as swiftly as possible in the opposite direction. Now he stands and observes, snapping pictures, mulling over the brief report that he will type on a PC at RLabs and send to his home on the blogspot:

> This is a scene all too familiar to me. I can remember the days that the cops used to pull us over after we went out to go and buy our drugs.
>
> In this incident the guys in the car were drinking in a public area. Because they were drunk, they were giving the cops a hard time.
>
> The cops called for reinforcement and did a search of the car. They also checked if the car was stolen.
>
> Thank you Lord that you have delivered me from all those foolish things. Where would I be without God? Things could have been so different.

In a separate blog, Mom 2.0, for 'women passionate about reconstructing their communities', we see a snapshot of a big, burly man with close-cropped hair, sitting next to a lady in grey, her hands hovering over the keys of a laptop, her gold-rimmed spectacles perched at the top of her nose. Brent Williams, the recovering addict, the lapsed gangster, is teaching his mother, Alida, how to blog.

Blogging, Marlon discovered, could be a way for people in troubled communities to channel their emotions and experiences, using the computer as a private confessional and a platform for public connection.

As part of his research into the transformative power of social networks, Marlon would ask his pilot group of bloggers at RLabs how they felt about being able to use this new medium of communication to tell their stories. 'Relieved and happy' was the recurring response.

The bloggers felt they could express their deepest feelings without being judged or feeling guilty, in spaces that belonged uniquely to them, and yet were open to anyone who wanted to wander in and look around. One addict-turned-blogger told Marlon he enjoyed blogging so much that it had almost become an addiction. But a good one.

Inside RLabs, light floods from a skylight onto an arrangement of workstations, each topped by a laptop or PC. It's warm as sunshine in here, and the quiet of concentration is broken only by the sharp, sudden tapping of keys, then silence, then a burst of tapping again.

It's the call-and-response of multiple conversations, conducted at a distance, from one screen to another, cutting straight to the point, no small talk, no chit-chat, no tiptoeing around the subject at hand.

i need support

how can i help

im using tik and desperately need help coz I have a 9 month

> *old baby*
>
> well we are here to help. how long have u been using?
>
> *6 years nd im only 18 nw*
>
> well are you willing to come and see someone in person?
>
> *yes i am. where*
>
> we are based in Bridgetown. 66 Tarentaal Rd. u free during the day?
>
> *yes thanx i'll come in next week*

Sitting at his laptop, Angelo King deftly clicks across to another conversation, one of more than 20 open in a series of mini-windows on his screen. A single word, from someone, somewhere, sending a message on a mobile phone. 'Hello.'

Angelo hits shift and taps a key.

> ?

A few seconds pass, and the answer flows.

> *can u help me. my sister is on drugs.*

Angelo flickers a smile. He's been doing this long enough to know that, sometimes, people seek help by proxy. They're asking for help for a brother, a sister, a cousin, a friend. It doesn't matter. He doesn't know their name, their location, their number. He just knows that they need the help. And he knows what that's like, because he's been there.

Angelo is an advisor on the Drug Advice Support programme at RLabs. He's 35, with a flash of orange in his spiky hair, and a slinky silver chain looped over his jersey. He's been free of drugs for two years, 'clean by the grace of God'. He was born on the Flats and grew up on the Flats, and he's made peace with the fact that you can't blame your environment for what you become. But you can change yourself.

Helo!;)

Hello how can i help u

My cuz is gng through a problem n luking 4 a place to help him

what is his problem?

His on drugs for 4 years already his also been in out of prison for a while at 1st it was just tik but nw its worst

how old is he?

Im nt sure bt i thnk hs 26

Tik. Tik-tik-tik. The drug is crystal methamphetamine, and it takes its name from the sound the rocks make when you heat them in a glass pipe, or a light bulb with the thread twisted off. The heat turns the rocks into a thick yellow vapour, which you inhale. It's a cheap and easy hit. It was Angelo's compulsion, and it turned him into a 'walking skeleton, physically, emotionally and spiritually shattered'.

> The stats at RLabs show that more than two-thirds of the requests for help revolve around substance abuse, and more than two-thirds of those are from people battling to get off tik.

The stats at RLabs show that more than two-thirds of the requests for help revolve around substance abuse, and more than two-thirds of those are from people battling to get off tik. Most of the advisors at RLabs are recovering addicts, trained by Lifeline in basic counselling techniques.

They listen, engage, inform, advise, and, where necessary, refer their clients to other organisations for professional help.

They're clients, or PSAs (persons seeking advice) – not victims or abusers or addicts. And wherever they are in the country, whatever their needs, there's one thing they tend to have in common. Access to a mobile phone, and Mxit.

Marlon was lecturing in the Faculty of Informatics and Computer Design at the Cape Peninsula University of Technology when he first saw Mxit in everyday use. His students showed him. He was struck not so much by the ease of instant chat, at next to no cost, as he was by the possibility that this could be a tool for teaching.

He helped to investigate and write a research paper on 'The usage of mobile instant messaging in tertiary education', using students aged between 18 and 23 as respondents.

> 'Being connected used to mean that you were reachable,' says Marlon. 'Now, being connected means that you want to share.'

The majority agreed or strongly agreed that Mxit could be useful for exchanging information, managing Q&A sessions and communicating assessment results. But the broken link in the chain was the faculty itself because Mxit was 'never or almost never' being used by lecturers. Clearly, concluded Marlon, this was a field for further research.

What the study had confirmed was the almost universal adoption of mobile as a communications technology by young people, and the rapid evolution of that technology beyond the voice-to-voice call. Voice was only third on the bouquet of preferred communications tools in the study, below SMS and email, and just above Mxit.

'Being connected used to mean that you were reachable,' says Marlon. 'Now, being connected means that you want to share.'

Marlon wanted to share. He went to see the principal of a high school on the Flats, and he said, 'Give me your ten worst pupils.' They would be the beta testers for a new model of peer-to-peer networking, connecting those who had learnt from their experiences with those who wanted to learn.

From that was born IDM Talk, a drug-advice-and-support service hosted by Impact Direct Ministries, and offering counselling by mobile chat for two hours after school every Tuesday and Thursday. It was a modest start, using the free Google Talk platform to connect advisors and their clients, with an administrator, or 'runner', managing the incoming requests and assigning them to advisors as they popped up on the screen.

That proved to be a messy proposition. With multiple advisors sharing a single GTalk account, and a moderator monitoring every conversation, the system was slow, insecure, prone to errors and confusion, and incapable of coping with more than a few dozen conversations a week.

But it led, as chaos often does, to a gathering of insightful minds, determined to apply a home-grown solution to a global problem. A Cape Flats solution.

It came out of RLabs, and today it is known as JamiiX, from the Swahili for 'community' and the 'X' for exchange – an elegant, scalable piece of software that allows many people to talk to many other people on social networks.

Technically, JamiiX is an aggregator and distributor of conversations, a single point of contact for managing multiple streams of information on the mobile internet and the Web. In practice, it is nothing more than a quiet, omnipresent facilitator that makes it easy for people to use technology to connect and to share. Software is simple. Real life can get complicated.

Hey

hey there

Hwu

well and you?

Nt gud

so whats happening?

I hv found out dat im 1mnth prgnt n its vry cmpctd

hw is it complicated?

My bf ex she prgnt n im also prgnt n hs fam wnt hm 2 marry de ex

ok so what happend with thorts of abortion?

We tlkd me n my bf he sd if we kp de bby hs fam mst nvr knw dat im prgnt n dat dy wl knw in de near future

bt hw do u feel abt keeping the baby?

Im vry scrd i wnt de bby n i dnt wnt de bby

are u scared of losing the boyfrend

No im scrd wat wl my fam sy n im also scrd 4 hm as hs fam wnts hm 2 marry de ex

hw old r u?

22 turng 23

During a typical two-hour session at RLabs, each advisor will have an average of 27 conversations to manage. Each conversation, on average, will consist of 29 messages. Each message, at the standard GPRS rate of two rand per megabyte of data, will cost the Mxit user approximately three cents, or 88 cents per counselling session.

A face-to-face session at a walk-in crisis or treatment centre, state or private, would cost about R200. But the advice and support on Mxit is not meant to replace professional help, says Marlon. The advisors at RLabs are first responders; they're like paramedics, rushing to the scene, assessing, assisting, performing triage.

The big difference is that the advisors do not see or hear their clients. They are worlds apart, gazing into separate screens, tapping staccato strings of text, waiting for the question or response that will keep the conversation flowing. And yet, they can touch lives.

> Technology is an enabler. It can uplift, empower, make a difference. It can give a glimmer of hope.

Technology is an enabler, says Marlon. It can uplift, empower, make a difference. It can give a glimmer of hope. This living laboratory in Bridgetown is a place of ideas and ideals, some of which, like JamiiX, become tangible products that can be exported across the world. Others, like Uusi, are works in progress.

Uusi. It's a Finnish word, meaning 'new'. As in new beginnings, new opportunities. It's a mobile social network, connecting job seekers and job placers. In its first few months, it attracted more than 100 000 registered users on Mxit. They uploaded some 60 000 mobile CVs and ran more than 2 million searches. It's based on an idea by Terence Hendricks, a counsellor and social media trainer at RLabs.

Terence also grew up on the Flats, in Manenberg. He dropped out of high school and struggled to find work. He remembers walking up and down streets, walking across empty lots, standing around, waiting, his hands idle and itching. There were many like him: an army of the unemployed. He wanted to be a tradesman, a fitter and turner. The closest he came was a part-time job, offloading and stacking tyres for Firestone.

He found his way, through the church, to RLabs. 'I felt like my brain was switching on,' he says. 'A whole new world of phones and computers. I had this vision of a space where people could use their skills, and use the technology, to help other people find work.'

He shared his idea with Marlon. Within months, it was a start-up, seeking funding and partners. Then it was a research paper, presented to a conference: 'The use of mobile technologies to address unemployment in the Western Cape'. Then it was an application on Mxit. Uusi. Your New Beginning.

Once, in Manenberg, Terence saw his father, a retired factory worker, walking around at the back of a mall. Not going anywhere, just wandering. It was that familiar shuffle of hopelessness, of joblessness, of nothing to do. Then he found a job: mopping the floor at a doctor's office. 'Now he can't stop talking about it,' says Terence. 'It has changed his world. Work gives you a purpose in life.'

Terence has a mobile phone in his hand, and he is thumbing through the listings and categories of job opportunities on Uusi. Business, finance and management; engineering and technology; services; computers …

He has a dream, too, of one day starting a business of his own. He is a little hesitant, a little shy, to say what it is. Then he shrugs. 'I do a little baking on the side,' he says. 'I haven't been taught how to bake, I just woke up one day and started baking. I baked bread. I baked muffins. I baked hot cross buns. I told my wife, bring me some recipes, man! I even made some chocolate mousse, although I don't eat it personally.'

He started taking orders from his wife's fellow workers at the university, and baked cakes and buns for special occasions. His dream, now, is to run his own coffee shop. In Simonstown, by the sea. He'll call it Toni's, after his little girl.

He remembers his biggest flop: a chocolate cake that somehow went wrong. He put it in the portable table-top oven at home, and it sank, a mess of sighing sponge and melted chocolate.

So he turned it all into a whole lot of cupcakes, and everyone was happy. He laughs at the memory. Here in the heart of the Flats, hope rises.

'M' IS FOR 'LITERATURE'

CHAPTER TEN

Mobile phones are for chatting, and books are for learning. But what happens when phone and book meet?

T HEY STOOD beneath the African sun, waiting, on the cusp
of winter, in queues that seemed to stretch back not just
hours, but years. They stood without much complaining, peer-
ing over heads now and then to see how far they had to go,
or looking over their shoulders to marvel at how far they had
travelled.

There was a buzz in the air, a murmur of anticipation, the
occasional hearty laugh at someone's joke. But mostly, they
just stood and waited their turn. They moved a step or a shuffle at a time, the long, slow march of history catching up with
the future. But there were signs, along the way, that the future
had already arrived.

There were people walking up and down the lines, at the tables with the cards and banners, in the official vehicles, talking
into hefty-looking devices, black slabs with big buttons and little telescopic antennae. It was 27 April 1994, and the wireless
radio cellular telephone was making its debut in South Africa.

In a country where less than 10 per cent of the population
had access to a fixed-line phone in their home, here at last was
a technology that would not only allow you to talk to someone,

but also do it while on the move, and let you call a number and reach the person rather than the place.

Or maybe these new devices would turn out to be nothing but 'very, very expensive toys ... about as necessary as the hula hoop, the yo-yo, and battery-operated luminous socks,' as a columnist for the *Cape Argus* predicted. It was too early to tell. The cellphone networks wouldn't be formally switched on for another month, but the national election monitors had been given a special dispensation to use the networks on the big day.

> Maybe these new devices would turn out to be nothing but 'very, very expensive toys ... about as necessary as the hula hoop, the yo-yo, and battery-operated luminous socks.'

And so they became symbols of the ringing in of social and political change, in a year that also saw the advent of a new and curious medium for accessing information on a personal computer. The World Wide Web, a graphical gateway to the infinitely hyperlinked wonders of the networks of networks, the internet.

Three freedoms in one year. The freedom to choose; the freedom to be heard; the freedom to know. But what of those born into these freedoms too young to have stood in the queue? The 'born frees', the 'mobile generation', the 'screenagers'.

For them, the downward gaze, the walk-and-text, the one-handed thumbing of a message to a friend, are actions as easy and as intuitive as breathing. The cellphone is not an accessory, it is a third hand, a second mind, a backup copy of your life in the clouds. What was life like before the phone? What would life be like without it?

Sitting in the lounge of her home in Somerset West, near Cape Town, Nina, a Mxit 'Superuser' and high-school student, regards the question with a sharp intake of breath, followed by a look of jaw-dropping bemusement.

'Very, very boring and depressing,' she says. And then, perhaps because her mom is sitting on the settee across from her, she adds: 'But I have control. I can put my phone away at any time.' And she puts her Samsung down for a few seconds.

She once spent nine hours on Mxit, straight, no break. 'Normally it's only five hours,' she hastens to add.

Her friends, Clara and Ané, also Mxit Superusers – meaning they have a couple of hundred contacts and spend more than 100 minutes a day on the network – nod enthusiastically, their phones still in their hands.

The thing is, during the school day, at least in class, in between breaks, you have no choice but to put your phone out of sight and out of reach. The girls claw inverted commas in the air.

'Cellphones are small devices, easy to slip behind a book or cradle under a desk, with the ringer on silent, and your eyes flitting from the screen to your teacher, the teacher to your screen. People are smart,' says Nina. 'It's not so much a compulsion, as a game. What's the worst that can happen?'

If you get caught using your phone in class, it'll be confiscated. You may not get it back for two months. Of course, you'll most likely have another phone at home, handed down by your mom or dad, left over from the last 24-month upgrade.

You can't not have a phone. It's unthinkable.

Even if you just leave your phone off for a few hours – maybe because you've fallen asleep while Mxiting – you'll have a flood of annoyed messages from your friends when you log back in, wanting to know: 'Why didn't you answer?' Nina rolls her eyes. 'If you want to send me a message, and I'm not on Mxit, if it's so important, SMS me. I mean, really.'

Clara says she hardly ever talks on her phone any more. She once spent nine hours on Mxit, straight, no break. 'Normally it's only five hours,' she hastens to add. What did she chat about, in all that time? 'Oh, nothing much,' she shrugs. But there are rituals, conventions, a template for kicking the conversation into touch.

'We just start by saying hi,' says Ané. Or 'aweh', a cross-cultural South African greeting that means the same thing, but with deeper philosophical undertones. Then follows a back-and-forth of enquiries, compressed into Mxit acronyms, designed to elicit information on what's happening, what happened, what you're doing, what you're planning to do, what you want to talk about, and what's news.

'Nobody can hear you on Mxit,' says Nina, 'and nobody else can see your messages. So we gossip a lot. But sometimes people save your messages and forward them to other people, so you have to be careful.' They prefer chatting to boys on Mxit, although they'll never accept an invitation from a boy they don't personally know. 'You just say, I don't know you, bye,' says Nina.

Her mom looks up, approvingly. 'They're good girls,' she says. 'I trust them.' She doesn't really understand them, though, and when they SMS, she often SMSes back to ask them to spell their acronyms and abbreviations in full.

Boys make better Mxit chat mates, says Nina, because they

tend to get your jokes and your little sarcasms. 'Girls can be so intense on Mxit,' says Ané. 'You say something sarcastic, and they're like, "Why are you being so mean to me!"' Boys, on the other hand, chatty and jokey on Mxit, often turn out to be shy and awkward when spoken to face to face. But there's more to the network than that.

You can learn from Mxit. You can play chess and other games. You can find out what you missed in class if you were absent. Teachers shouldn't complain about kids using Mxit, argues Nina. 'If they really want to know what's going on in our lives, they should go on Mxit themselves, and find out.' Would she be okay with that? A look of horror in her eyes.

'I just wouldn't add them as a contact,' she says. 'It would worry me that they had all that time to go on Mxit.'

But here is a teacher who does go on Mxit. Her name is Ros Haden, and she taught at St Mark's College, a co-educational school in Limpopo. She has bright eyes, a cheerful face, and a soft voice that makes you sit up and listen.

Today her place of learning is Harmony High, which you will find in the township of ... well, you won't find it. It is a fictional school, the setting for a series of novellas aimed at teenage readers, who can read the books, for free, on Mxit. They are commissioned and distributed by the FunDza Literacy Trust, based in Cape Town. FunDza means to read or to learn, in a fun way, in ZA.

This is a school of literature – short books written and formatted for the mobile phone – that is variously known as M-Lit, or MxLit, or mo-books, or mobi-books, or m-books. But the point is, they're books, to be read a screen at a time, a chapter a day, with each chapter ending on a cliffhanger that leaves you itching to download the next one.

'Here.' Her mother handed Ntombi a five-rand coin from her new gold bag. She smelled of some strong perfume Zakes had bought her. 'Buy yourself some sweets at the shop,' she said as she rushed out, putting on lipstick as she went.

'Mama, I'm meant to be at singing practice. The competition is next week and ...' But her mother was already out of the door and in the seat of Zakes's resprayed BMW with its fluffy dice bouncing from the rear-view mirror and couldn't hear her. All she could do was watch as Zakes reversed with a squeal of tyres, and then they were gone.

Thus ends the first chapter in the first book of the Harmony High series, *Broken Promises*, by Ros Haden. Readers are urged to give their feedback via Mxit, and they do. Says Tebogo: 'Mama ntombi lvs hr kids bt is blindd by lv nd al material thngs dat zakes cn prvide unaware dat shes misin owt on hr daugtrz lives nd poor ntombi shes gt 2 b a mada 2 zinzi nd a chld on hr own.'

Says yellobric: 'Don't like this Zakes guy at all; although fluffy dice are the bomb.'

For the screenage generation, accustomed to viewing the world through the screen of a mobile phone, these concise stories, gritty, fast-paced and dramatic, serve not just as a window, but as a mirror.

'We try to make the stories authentic,' says Ros. 'Positive, but not preachy. Our primary aim is to popularise reading, to the point where the readers are so hooked by the story, that they don't even realise that they're reading.'

Ros draws on her experience of working with teenagers, helping them to deal with their issues. 'And of course,' she says, 'I was a teenager myself, once.' But FunDza also acts as a platform for teenagers to tell their own stories, and one of the most popular mobi-books is *Nobody Will Ever Kill Me*, a personal narrative of township life by Mbu Maloni, aged nineteen.

> *What does one remember first in life? Maybe a smell, a voice, the warmth of the skin of your mom? I try to remember and I can't. The first thing I remember is something I am ashamed of. Maybe that's why I don't want to remember. I want to forget. I don't have a first nice memory in life.*

The book has a raw texture, a tang of blood and tears, and the text on a handheld screen gives it the intimacy of a confessional, a private message sent on Mxit.

'Our readers talk about old school,' says Mignon Hardie, 'and by old school, they mean a printed book. New school is an e-book or a mobi-book. But we're starting to see a lot of readers, once they've read the new-school version, asking if they can have a copy of the old-school version too.'

The first 'IM book' on Mxit – instant mobile, as in instantly downloadable on Mxit – was *Emily and the Battle of the Veil*, a tale of 'teenage spiritual fantasy fiction' by Karen Michelle Brooks. It opens with our heroine, Emily, thirteen, stretching her arms above her head after spending most of the afternoon escaping into unknown and exciting worlds … by reading a book.

Karen, with a degree in psychology and a background in the IT business, published her debut novel herself, after being told by publishers that there wasn't a market for teenage fiction in South Africa. But the print quality of her book was so poor, falling apart at the spine, that she started looking for a less messy means of distribution.

Mxit liked her proposal, and her book was published on the network in 2009 as a series of downloadable chapters, for R13.50 each. Her first two Emily tales, in total, have so far sold more than 100 000 chapters. Proof that there is moola to be made from putting a book into a phone.

In the same year, Mxit published – or 'm-published' – a 21-chapter teenage mystery novel called *Kontax*, by Cape Town author Sam Wilson. Aimed at readers aged between eleven and eighteen, the book was translated into Xhosa, and was used as the axis of a study into mobile literacy ('m4Lit') in two townships, Langa and Gugulethu.

Central to the study was an apparent paradox, outlined by the report's author, Marion Walton, of the Centre for Film and Media Studies at the University of Cape Town.

> The book has a raw texture, a tang of blood and tears, and the text on a handheld screen gives it the intimacy of a confessional, a private message sent on Mxit.

'In most of the country's underperforming schools,' she wrote, 'a majority of teens are left behind academically. Many experience difficulties with literacy instruction and most have limited access to books and computers.'

At the same time, thanks to the mobile-phone revolution and a thriving mobile youth culture, many of these teenagers interact richly with the written word on social networks such as Mxit. Marion interviewed 61 Xhosa-speaking youths in the low-income focus areas, where 100 per cent of teenagers have access to mobile phones, either their own or shared with family or friends.

More than 63 000 Mxit users signed up to read the *Kontax* book, downloadable chapter by chapter, with about 7 000 downloading all the chapters. Marion found a high level of enthusiasm for mobile reading among her survey participants, suggesting that the mobile platform may be a useful tool for encouraging reading and writing outside the formal school arena.

> This is a generation at ease and at home with mobile technology, adept at distributing their attention across multiple chats on Mxit, and juggling chatting with their social and academic obligations.

Mxit was by far the predominant social network used by the group, with more than 95 per cent having used it at some point.

'While they present themselves as adept social navigators,' writes Marion, 'the teens did admit their difficulties with managing Mxit as an all-consuming, always available link to a world of perpetual social interaction.'

Many set strict rules for themselves, limiting their access to the network during homework and study time. One participant outlined his routine: 'I study from eight to nine, there's a phone, there's a book. Immediately when I get a message I leave the book, and attend to the phone and chat, and as soon as I put down the phone I get another message. So the problem is I just close the book and don't study.' The battle between phone and book would continue throughout the night, with a rest from both pursuits between 3 am and 4 am, before he set off to school at 6 am to write an exam.

Staying up late at night was identified as the chief cause of 'inter-generational conflict about cellphones and Mxit' by the focus group, with 'keeping bad company' and 'incorrect grammar and spelling' at the foot of the list.

This is a generation at ease and at home with mobile technology, adept at distributing their attention across multiple chats on Mxit, and juggling chatting with their social and academic obligations. For all the potential sources of conflict with parents and teachers, Mxit use has given rise to an 'explosion of written interaction among teenagers', perhaps paving the way for a more equitable and beneficial relationship between book and phone. And teachers, it seems, are learning too.

Marion tells the tale of a group of students chatting on Mxit during school, when one of them receives an invitation from a new Mxit contact, who appears to be a teen. They subsequently discover, to their horror, that their new contact is their teacher, Miss Cupido.

One way, perhaps, in this new age of mobile chatting and learning, for the older generation to keep the younger in line.

THE HUB OF THE HUMAN HEARTBEAT

CHAPTER ELEVEN

Online social networks are circles within circles, sending out signals that can cross boundaries and bring people together in the real world.

THE MAP of Africa is glowing. Soft pulses of light, radiating from the subcontinent, spreading like ripples in a pond. Some are as faint as watermarks; others burn with a fiery intensity. From afar they look like clusters, hot zones of thermal activity.

Zoom in to the streets of the cities, the townships, the suburbs, and the clusters explode and subdivide, spawning discrete cells, pulsing and fading, pulsing and fading.

Here is one in Sam Nujoma Avenue in Windhoek, Namibia. Here is another in Jason Moyo Avenue in Gweru, Zimbabwe. And here, too, in Kingsway Avenue, Maseru, Lesotho. In the bigger centres, Cape Town, Johannesburg, Durban, there are many of them, a constellation of blips, reflecting and absorbing each other's energy. The map is alive.

We are looking at Mxit in Motion, a real-time geographical display, on a Web browser, of the traffic of connections on an online social network. Each circle of light represents a completed Mxit user session, a successful negotiation between an external device and our servers in Stellenbosch.

It is tempting, in the cold light of analytics, to see these circles as hubs or nodes of data being sent and received. But they are something more than that. They are people, in conversation with each other.

In Douglas in the Northern Cape, Thomas, a court stenographer – 'the DJ of the court', as some call him – unwinds with a movie after a busy day of interesting and not-so-interesting cases. Just after 10.30 pm, he logs onto Mxit for a chat with family and friends, 'my chance to say hi to people I care about'.

In Port Elizabeth in the Eastern Cape, Vuyolwethu, a university student, takes a break from her assignment, reaches for her cellphone, and taps into her MPCB: 'My Personal Communication Broadcaster'. She catches up with news and gossip, plays a game and downloads a song, all on Mxit.

In KwaNobuhle, Uitenhage, bow.weedy3, one of six siblings, wonders what it will take to make his dream come true. He wants to be an actor and a hip-hop dancer, starring in a movie he's written about the people he's met on Mxit.

We know the numbers, the click-click-click of the counter advancing, the 50 million in more than 120 countries, the 10 million active users in South Africa, the 750 million messages sent every day. We know that a quarter of our users are aged between 13 and 17, and almost half are 18 to 25, and 10 per cent – that's a million users – are 35 and older. We know that up to 50 000 new users register every day.

We like to think of Mxit as a community, but it is really a multitude of communities, with each user constructing a mini-network of their own, by inviting and adding contacts and being added in return. Circles within circles within circles.

Remember 2010? In South Africa, it was the year of the Beautiful Game. Flags flying from car windows, commuters wearing football shirts to work, the raucous call-and-response of vuvuzelas filling the air. And through it all, rising from the streets, the refrain of a song that became an unofficial anthem, a catchphrase, a call to arms and feet and hips: 'Make the circle bigger, bigger, bigger! Make the circle bigger!' Who could resist an invitation like that?

We are by nature social animals, driven by the urge to be part of something bigger than ourselves – a family, a clan, a community, a nation. That same instinct is unlocked in on-line social networks, where the big difference is one of proximity. We are alone together, us and our machines, not talking but chatting, not dancing but tap-tap-tapping. But this mode of communication brings an interesting dynamic to the party.

> We are alone together, us and our machines, not talking but chatting, not dancing but tap-tap-tapping. But this mode of communication brings an interesting dynamic to the party.

On Mxit, when you step inside the circle, when you make the circle bigger, you can step outside of yourself. You can be a different version of who you are, or you can be the someone else you have always wanted to be. Anonymity is hard-wired into the superstructure of the network.

You build a circle of trust around the people you know, the people you talk to one on one. Beyond that, in the buzz and babble of the virtual chat rooms, where up to ten people at a time convene for casual conversation, flirting, or verbal joust-ing, you protect the integrity of your identity with a username of your own choosing. It can be playful or enigmatic, revealing or misleading. It can be a tangle of emoticons and punctuation marks, or a plain name with a suffix of digits to set you apart from all the other plain names. Either way, nobody needs to know who you really are.

There are good reasons for wearing a mask online. You may want to escape for a while from reality and the conventions of social discourse. You may want to engage in idle banter with a bunch of friendly strangers, without the risk of commitment or consequences.

You can be a different version of who you are, or you can be the someone else you have always wanted to be. Anonymity is hard-wired into the superstructure of the network.

You may want to express candid, provocative opinions that would be out of place in the workplace. You may want to shelter yourself from being hard-sold a product on the basis of your social profile. Or maybe you want to pretend to be someone else, in the hope that you will discover something new about yourself. Anonymity can set you free.

Let us consider the case of HappyMoments. She is in her 40s, and she lives in a suburb on the Cape Flats, near Cape Town. Her real name is...well, that would be telling. But she is HappyMoments in the chat rooms, even though HappyMoments isn't really who she is, and she wears her alter ego with good grace and the best of intentions.

Her favourite chat room on Mxit is Room 6 on 30 Something. It is one of the most popular rooms on the network, and it really comes alive after working hours, when the thirty-somethings pop in to take the weight off their feet, trade hellos and innu-endos, and enjoy a good moan about the weight of the world.

Just as in the real world, where rooms have real walls and doors, the tone of conversation can quickly shift from cordial to tetchy, and cutting remarks can be misinterpreted even when de-fanged by a smiley. That's when HappyMoments steps in.

She is a soother of tensions, shining a light on the bright side, quick with a quip to defuse a petty squabble. And her chat room-mates listen and learn, because she has wisdom be-yond their years. HappyMoments, from the moment she steps into the room, becomes someone else: a 60-year-old domestic housekeeper from the Flats, working for a difficult employer in the wealthy Winelands village of Franschhoek.

> Just as in the real world, the tone of conversation can quickly shift from cordial to tetchy, and cutting remarks can be misinterpreted even when de-fanged by a smiley.

She chose that persona primarily be-cause she didn't want to be 'hit on' by the guys in the room, but also because it was close enough to home for her to make it convincing. She re-ally is a hard-working housekeeper, at least in her own house, and when she com-plains, tongue-in-cheek, about her pushy madam, well, she re-ally has her husband in mind.

She sees herself as playing a role, rather than pretend-ing; she is not stretching the truth as much as mixing it up with a little fun. She sees people arguing and fighting in the chat room, and her instinct is to make them smile. And if that doesn't work? 'It's not real life,' she says. 'It's a chat room. If you don't like it, you can leave. You have a choice.'

She draws a line between the online world, with its friends who are really strangers, and the real world, where you can sit opposite someone and talk to them face to face, without having to resort to emoticons to get your point across. She has crossed that line many times.

It was her daughter, then a teenager, who introduced her to Mxit in 2006, when the network was spreading like a veld fire across the Cape Flats. She thought her mom would enjoy the chatting, and that it would help her

> 'It's not real life,' she says. 'It's a chat room. If you don't like it, you can leave. You have a choice.'

burn some of her excess energy, a restlessness that came from being retrenched after more than 20 years in the same job. But she also had some advice: 'Be careful, Mom. Mxit can eat you alive if you don't look out.'

And so, under the guidance and mentorship of her daughter, HappyMoments made herself at home on Mxit, observing all the rules – don't be too quick to trust people, don't give out your PIN, real name, or contact information – and setting strict boundaries and conditions, including the strategic use of the 'dot ignore' command.

If you .ignore a user in a chat room, you will no longer see their messages, and HappyMoments would do this with all the regulars she met on 30 Something, discreetly taking a break from their company after two weeks.

'The people you meet are there for all sorts of reasons,' she says. 'Some are there to play, some to hurt. I didn't want to get emotionally involved. You can't rush to trust people, or see them as your friends. There's too much pretending.'

But then, as she herself became a regular, she began to see through the pretence, getting glimpses into the lives of people who were just like her – the real her, not the 60-year-old house-keeper in Franschhoek. Other women, too, were struggling to get by, struggling to find work, struggling to look after their families and educate their children.

Slowly, carefully, she began using another command: .gpm, to send a private message to a user in a chat room. She would offer help, encouragement and advice, still in the guise of HappyMoments. She became a sort of agony auntie on 30 Something, someone you could talk to and trust, and people would 'dot gift' her with small amounts of airtime as a token of thanks. (You can also .hug or .tickle someone in a chat room, or send them a .slap or a .bomb if they're annoying.)

Finally, HappyMoments felt comfortable enough to break the cardinal rule. She began meeting her chat room buddies IRL: In Real Life. She wanted to stop running away from her own identity, even if it meant breaking the barriers she had set up. She wanted to connect with people, to prove that we are all connected, that 'we are who we are because of each other', as the African ideal of ubuntu puts it.

She reckons she has met about 50 of her chat room bud-dies in person, and although there have been a few 'rotten apples', her instincts have largely proved to be right. 'You can tell who the good people are,' she says.

Today she runs a small business, a micro-entrepreneur-ship, from her laptop and her phone, acting as a facilitator, a finder of jobs and opportunities, a marketer of other people's micro-businesses – a cake-crafting shop, a design studio, a beauty salon – using her blog and Facebook and Twitter pages. She has about 100 people in her database, and she charges just enough to cover her costs and make a small profit when business is good. But it's more than a business. It's a social network. HappyMoments has made the circle bigger.

In the real world – the analogue world, where people make connections face to face – we don't count our friends. We count on them, if we are lucky enough, when the need arises, but we don't accumulate and categorise them as Friends or Followers or Contacts or Connections.

In the digital world, we keep a tally, and we feel the pain, for real, when we are unfriended or unfollowed or deleted from the list. But why do we join online social networks in the first place?

Sometimes it's to forge new bonds or strengthen existing connections of friendship and family and workplace and school: Facebook. Sometimes it's to stand on a platform and broadcast our thoughts, experiences and observations to the world: Twitter.

Sometimes, it's because we want to showcase our better selves in the hope that the right people take notice: LinkedIn. And sometimes it's simply because we are higher primates with an instinctive and compelling urge to hang out with the pack: CafeMom, for mothers and mothers-to-be; TravBuddy, for travellers heading to the same location; Epernicus, for scientists working on research projects.

> On the internet, we are not just social animals, we are polysocial animals, hopping from network to network until we find the one that most feels like home.

There are dozens of online social networks, offering dozens of opportunities to connect, to share, to belong. Each network needs an incentive of association, a 'kinship proposition' that will make us feel we are in the right space, with the right people. (We don't need to like them, or be of like mind, but we do need to like networking with them on some level.)

On the internet, we are not just social animals, we are polysocial animals, hopping from network to network until we find the one that most feels like home. One afternoon in Stellenbosch, Professor Wynand Coetzer is sitting beneath the oaks, at a pavement café, stirring his cappuccino, when a car eases out of a tight parking space and another car judders to a halt and nudges in, the driver and passenger whooping with joy. The Prof. winces. This place isn't what it used to be.

The property developers are destroying it building by building; once you get out of the old town centre, you could be anywhere, in the anonymous sprawl of cluster estates and mini-malls and office parks. And just getting in and out of Stellenbosch these days is a nightmare.

But Professor Coetzer has been here for most of his life,

teaching electrical engineering at the university, getting to grips with the evolution of machines, from valves to transistors to mainframes to microprocessors to mobiles.

Today he teaches electrical engineers how to be entrepreneurs, and he offers advice to new businesses in this town of tech start-ups. And he networks. In the real world, and online.

'In a way,' he says, 'our real social networks are also virtual, because you can't see them or trace them. They are based on your perception of other people, and some kind of subconscious response to other people, and their response in turn. And then, somehow, you decide you're going to have a braai together. It's actually very virtual, but also very real.'

He remembers being called in a few years ago, to consult with a small tech start-up run by a former student at Stellenbosch. A guy named Herman Heunis. Herman and his team were throwing around ideas for the company – mobile games, a mobile search engine, mobile classified ads – and the Prof. was shooting them all down.

'And when I opened my eyes again,' he says, 'they had Mxit.'

He always tells start-up entrepreneurs that it is going to be very difficult to break into a new market, largely because of our natural aversion to risk, our reluctance to try something we haven't tried before. But as soon as we have a point of reference, as soon as we can say 'that guy's similar to me, and it seems to be working for him', the risk tends to dissipate. That is the power of the social network.

If your friends are on Mxit, he says, then you're going to get onto Mxit. You won't give a damn about the network and how it works. You will care about the image and the proposition, and on Mxit, that proposition is as simple as human nature itself. 'The message of Mxit,' says the Prof., miming the jabbing of thumbs on the keypad of a cellphone, 'is to message.'

This is the story of Da Twista and Da Storm. They met one morning in a chat room on Mxit. In between the clicking of keys, by two total strangers, thousands of kilometres apart, something clicked. They took their conversation to a private space in the room, and they learnt a few things about each other.

Da Twista was Matt, a civil engineer from East London. Da Storm was Sabeneth, Sabie to her friends, a student from the little town of Moorreesburg in the Western Cape. They would chat for hours, talking about work, study, the little details of their separate lives. One evening, when they had known each other for about six months, Matt made a proposal. 'Why don't you come down to East London for a couple of days?' he asked.

Sabie was a bit sceptical, a bit nervous, but at the same time she thought to herself, 'If you don't go, you'll never know, and then you'll always be wondering.' In January 2006, she caught the plane from Cape Town, and Matt was waiting at the terminal. They had swapped photographs over Mxit, so they knew who to look out for.

'To my surprise,' recalls Matt, 'when I saw her, she glanced back and walked right past me.'

Then she stopped. Matt would discover that Da Storm, so outgoing and chatty online, was very shy in real life. Sabie thought that about Matt too. But she also thought 'wow!'

Today Matt and Sabie, Da Twista and Da Storm, are Matt and Sabie Kamffer. They live in Stellenbosch, and they still talk to each other on Mxit.

The circles on the map, softly pulsing, sending out their signals from person to person, from street to street, are the symbols of a social network in constant motion. But they're also something else. They're human heartbeats, seeking a connection.

One hundred trillion waves of freedom

CHAPTER TWELVE

A sleepless night, a walk on the beachfront, and a hot-footed epiphany that brings the true value of a twenty-first-century African social network to light.

BEFORE me lies the bed of burning coals, a narrow pathway fringed by tufts of freshly planted grass. The leaping flames have been brought under control, the hot rocks patted down and put in their place. I can feel the heat of the embers rising through the shimmer of smoke. No problem. I'm a South African, instinctively unafraid of smouldering coals, tough enough to hold my hand above them unflinchingly for ten seconds, when the time is right to braai, even though I'm a *soutie*. But these are my feet we're talking about. My bare feet.

I flex my toes. I stand up straight. I take a deep breath. It's time to walk the talk. I can see them watching me out of the corner of my eye. The Mad Hatter, with his tartan trousers, his frilly shirt and his orange curls exploding around his ears.

The Queen's Guards, with their poker faces and their spears tipped with hearts and diamonds and clubs and spades. Alice, with her pinafore dress and her sensible shoes. Tweedledum and Tweedledee, with their striped jumpers and red suspenders.

Me, I'm wearing my everyday office clothes. Jeans and a T-shirt. Usually, I slip on a pair of slops too, but I'm about to walk on fire, and it's a Sunday. The first day of April. A red-letter day on the calendar, for two reasons.

It's April Fool's Day, the day on which the credulous and the gullible are easily swayed, and you can't believe anything you read in the papers. Then, by sheer coincidence or grand design, 1 April is also the second anniversary of World of Avatar, the original manifestation of my dream.

A good day to start a business, or take stock of a business you've started. In my case, Mxit, the biggest social network in Africa.

Now here we are, exactly two years after the beginning of this story, having an Alice in Wonderland party on a wine farm in Stellenbosch, to mark the official launch of Mxit 2.0.

We've come a long way. An extra six or seven steps, barefoot on a bed of burning coals, isn't going to hurt.

> I'm about to walk on fire, and it's a Sunday. The first day of April. A red-letter day on the calendar, for two reasons.

I think back to a couple of nights ago. I'm in an apartment in Cape Town, with Sibella and the girls. I can't sleep. I haven't slept properly in days.

I lie in bed reading *Game of Thrones*. It's a dark saga of medieval lords and warriors and castles, and a king who sits on a throne made of swords, fused into shape by the breath of a dragon. At 10.30 pm, I feel my eyelids getting heavy. I hit the light switch, start to say a prayer of thanks and fall asleep halfway through.

Then my eyes open wide, jolted by a dream that drifts into the haze. I glance at the clock. It's 2 am. The numbers are spinning through my head again. We're burning about R4 million in cash at Mxit every month. We have about R80 million in hand. That means we have 18 months and then ... stop. Don't panic, as I like to say in my emails and my updates to stakeholders.

We're on the right path. We need to grit our teeth, focus and carry on walking. We can't let ourselves fall into the trap of letting short-term cash flow dictate long-term strategy.

> Every organisation grapples with cognitive dissonance, the tug of war between simultaneously held, diametrically opposing beliefs and ideals.

Every organisation grapples with cognitive dissonance, the tug of war between simultaneously held, diametrically opposing beliefs and ideals. At Mxit, the example that springs quickest to mind is the PIN reset.

It's a cumbersome, unfriendly process, and it costs the user R2 via a premium-rated SMS. We could probably fix it. We could probably make it friendlier. At the same time, it generates R2 million in revenue for Mxit every year. So it's in our interests to keep the system as it is. But is it in the best interests of our users?

We need to think bigger, see further. Not just into the stars, but into our own hearts. A business does not make money. It makes product. Money is the result of the product. But no one is in it for product alone. We're in it for the purpose. If we're doing what we're doing just to make money, then we're dead. If we're doing it for ourselves, then we're foolish as well as selfish.

So why are we doing it? It's two in the morning and I'm going for a walk. I get out of bed, put on my jeans, T-shirt, jacket and slops, and leave the apartment, emerging onto the Sea Point promenade.

The Atlantic Ocean is crashing on the rocks below, quieting the crazy noise of my thoughts. The promenade, usually buzzing with the traffic of dog-walkers and joggers and laughing, playing children, is almost completely deserted.

I watch the waves for a while, the way they curl out of the black water, advancing, retreating, regrouping, forever trying to reclaim the shore. Now I'm thinking about the numbers again.

One of the big attractions of Mxit, as an investment proposition, was the notion that it could evolve into something more than a communications platform, more than a social network. It could become a platform for any kind of digital content. Airtime, electricity, music, games, anything that is invisible.

> If we're doing what we're doing just to make money, then we're dead. If we're doing it for ourselves, then we're foolish as well as selfish.

In South Africa, the average prepaid mobile user spends R100 a month on airtime. So if Mxit was to attract just one million users – 10 per cent of its active base – to a telco of its own, and retain 30 per cent of that monthly airtime spend … we'd be looking at an extra R30 million a month. That's something worth chasing.

I want to build something that helps other people make money. Build an ecosystem of people that are getting ahead.

That's the kind of thing that can go viral. Suddenly, we're exponential. There is a theory of computer networking called Metcalfe's Law, named after Robert Metcalfe, an American electrical engineer, who was one of the inventors of Ethernet technology.

Metcalfe's Law says that the value of a telecommunications network is proportional to the square of the number of connected users of the system. This is a lot simpler – and, I'm afraid, a lot more complex – than it sounds. All it really means is that the value, or utility, of a network increases in proportion to the number of nodes on the network.

> The greater the number of users, the greater the value of the network because each user has the potential to connect with and attract other users to the network. So it grows.

A node can be a computer or a telephone, or, for our purposes, a human being. A Mxit user. So let's say we have ten users on the network. Their value, squared, would be 100. But if we have 100 users, the value of the network doesn't just go up tenfold, it goes up by a factor of itself, to become 100 squared, or 10 000.

The greater the number of users, the greater the value of the network because each user has the potential to connect with and attract other users to the network. So it grows.

If we accept the modest proposition that Mxit has 10 million active users, then the value of the network, according to Metcalfe's Law, is 100 000 000 000 000. One hundred trillion. I don't even know what that means.

When numbers get that serious, they slip from your grasp,

floating into the ether like bubbles. There's no point trying to pin them down. The real value of Metcalfe's Law, of the network effect, does not lie in the numbers. It lies in the value.

So the big question we need to ask is, what is the value of our social network? What is the value of the technology that facilitates and sustains it? What is the real, lasting value of Mxit?

I've been strolling on the promenade for a couple of hours. The colour of the sky is changing. The waves are crested with a tinge of fiery light. The first joggers are foot-falling their way along the concrete. I sit on a bench and look at the ocean for a while. Then, slowly, the answer dawns on me.

We're doing this because ... but wait. Who are we, anyway?

In the valley of the vineyards, in the cradle of the mountains, I've assembled a team of warriors, wizards, sorcerers and bravehearts. Avatars. I'm asking them to go into battle with me, to cross the blazing coals, to venture into unknown worlds. I'm asking them not to panic. I'm asking them to believe.

Our battleground is the fastest-changing segment in the fastest-changing sector in the world. Social networks on the mobile internet. We need, first of all, to shine a light on our own network, our own small band. We don't have time to worry about those who aren't game, who hide behind rules and regulations. Delete the rules and regulations. Give everyone a chance to steal or mess up. Find the bad apples early, and move on quickly.

> In the valley of the vineyards, in the cradle of the mountains, I've assembled a team of warriors, wizards, sorcerers and bravehearts.

Then we need to build trust. Not buddy–buddy trust. Crazy

banshee war trust. The kind of trust where you know you won't be alone when you're facing an army of slavering sociopaths armed with sharp knives and no scruples.

We're building a social network, not a pyramid. Each node in the network is the node that makes the other nodes work. Each node is the spark that sets the magic free.

Freedom. That's why we're doing this.

The freedom to communicate.

The freedom to connect.

The freedom to enlighten.

The freedom to learn.

The freedom to be anonymous.

The freedom to work.

The freedom to share.

The freedom to change.

That's the real value of these blocks of silicon and plastic and circuitry that we hold in our hands. That's the real value of a mobile social network in Africa. That's the value of Mxit: one hundred trillion waves of freedom. It's April Fool's Day. The sun is shining on the vineyards. There's music in the air.

I've told everyone – the Mad Hatter, the Queen's Guards, Alice, the Dormouse, the Tweedle Twins, the White Rabbits, the Hubbly-Bubbly Smokers, the rest of the bunch in their jeans and T-shirts – about our plans for the network, about our cashless payment apps, our telco, our partnerships with businesses and governments, our community outreach programmes. I've told everyone what we're doing to make money, and how we're going to help other people make the money that will make us more money in turn.

And now, I'm taking a deep breath and I'm walking on fire.

One, two, three, four, five, six, seven. It looks like magic. It looks impossible. But like so many small miracles we come to take for granted, it's nothing more than a matter of belief.

If you walk briskly enough across a bed of hot coals, the heat won't have time to penetrate the thick epidermis of your soles. And if you tell yourself you're not walking on hot coals, you're walking on cool grass, then cool grass is what it will feel like. And you won't get burnt.

Freedom from fear. I think that's the most important thing. What's there to be afraid of? Death? Well, you won't have anything to worry about when you're dead. Failure? Well, if you fail, at least you're not dead. Are you afraid of what other people might think? Let them think. If you're doing it for the right reasons, you have nothing to fear.

Me, I've walked on fire on April Fool's Day. No big deal. I started a business on April Fool's Day. Slightly bigger deal. So the only thing I really have to fear is that in the midst of it all, I might forget my family. If I ever reach an inflection point where I must choose, I will have failed. I can't allow that to happen. And I can't leave an army on the field again.

For now, freedom means opportunity. We've got the people, we've got the technology, we've got the network. We've got the right moment, in the right place, on the right continent. We've got all the ingredients we need to make this formula work, and we're ready to Mxit.

THE
MXTIONARY

*A glossary of the
Mxlish language*

1CE	*Once. Also, 1nc*
2day	*Today*
2geda	*Together*
2mor	*Tomorrow*
2U2	*To you too*
4ever	*Forever*
4get	*Forget*
4giv	*Forgive*
4me	*For me*
4rm	*From*
4frnd	*Friend*
4wndz	*Friends*
Abt	*About*
Ada	*Other (as in, 'da ada day')*
AFAIC	*As far as I'm concerned*
AFAICS	*As far as I can see*
AFAICT	*As far as I can tell*
AFAIK	*As far as I know*
Afta	*After*

Agn	*Again*
Alt	*A lot*
Anada	*Another*
Ansa	*Answer*
ASL	*Age, sex, location? (Mostly used in Mxit chat rooms, as an opening conversational gambit. Frowned upon and even banned in some chat rooms, where it is regarded as intrusive and irrelevant.)*
ASLR	*Age, sex, location, race?*
ATB	*All the best*
ATM	*At the moment*
Awsum	*Awesome*
B2wn	*Between*
B4	*Before*
BB4N	*Bye-bye for now*
BBL	*Be back later*
BBN	*Bye-bye now*
BBS	*Be back soon*
BCNU	*Be seeing you*
BCOZ *or* **BCZ**	*Because*
Bdae	*Birthday*
BF	*Boyfriend*
BFF	*Best friends forever*
Blv	*Believe*
BRB	*Be right back*
Bt	*But*
BTW	*By the way*
C	*See*
Col	*Call*

CU	*See you*
Cud	*Could*
CUL8TR	*See you later*
CYA	*See ya*
D8	*Date*
Da	*The (pronounced 'the', not 'da')*
Da1	*The one*
Dae	*Day*
Daez	*Days (The daez of the week on Mxit are Mndae, Tsdae, Wnsdae, Thrsdae, Frdae, Saterdae, Sndae.)*
Dat	*That*
DC	*Disconnected*
Ddnt	*Didn't*
DILLIGAD	*Does it look like I give a damn?*
Diz	*This*
DL	*Download*
Dnt	*Don't*
Drmz	*Dreams*
Enuf	*Enough*
Every1	*Everyone*
Evrydng	*Everything*
Ez	*Easy*
F	*If*
Fingz	*Things*
Fn	*Fine*
Frnd	*Friend*
FTW	*For the win*
Fwn	*From*

Fwnd	*Friend*
Fynd	*Find*
G2G	*Got to go*
GAL	*Get a life*
GF	*Girlfriend*
GL	*Good luck*
Gr8	*Great*
GTG	*Got to go*
Gud	*Good*
Gv	*Give*
Gyz	*Guys*
H8	*Hate*
HB	*Hurry back*
Helo	*Hello*
Hez	*He's*
Hlp	*Help*
Holidaez	*Holidays*
HOT4U	*Hot for you*
Hpy	*Happy. Also, hapi*
Hu	*Who*
Hud	*How you doing?*
HUGZ	*Hugs*
Hy	*Hi*
IC	*I see*
ICYDK	*In case you didn't know*
ID10T	*Idiot*
IDC	*I don't care*
IDK	*I don't know*
IM	*Instant messaging*

IMHO	*In my honest/humble opinion*
IMO	*In my opinion*
In2	*Into*
IOW	*In other words*
Iz	*Is*
JIC	*Just in case*
JJ	*Just joking*
JK	*Just kidding*
JOO	*You*
JT	*Just teasing*
JW	*Just wondering, or just wondered*
K	*Okay*
kewl	*Cool. Also, kwl*
Knw	*Know*
L8	*Late*
L8ER	*Later. Also, l8r or l8tr*
LMAO	*Laughing my ass off*
LMSAO	*Laughing my sexy ass off*
LOL	*Laughing out loud*
LQTM	*Laughing quietly to myself*
Luk	*Look*
Lulz	*Laughs*
Lv	*Love*
Lyf	*Life*
Lyk	*Like*
M8	*Mate*
Ma	*My*
Maself	*Myself*
Mi	*Me*

Mit	*Meet*
Mo	*Moment*
Moola	*The Mxit currency, based on airtime*
Mrng	*Morning*
MSG	*Message*
MYOB	*Mind your own business*
NBD	*No big deal*
Nd	*And*
NE	*Any*
NE1	*Anyone*
NETHNG	*Anything*
Neva	*Never*
NM	*Not much*
No1	*No one*
NOOB	*Newbie, as in one who is unacquainted with the ways of Mxit*
NP	*No problem*
NRG	*Energy*
Nt	*Not*
Nthng	*Nothing*
NUFF	*Enough said*
NVM	*Never mind*
Nvr	*Never*
Nw	*Now*
NW	*No way*
Nxt	*Next*
Nym	*Name*
Nymd	*Named*
Nyt	*Night*

OBTW	*Oh, by the way*
OIC	*Oh I see*
Ova	*Over*
Pc	*Peace*
PITA	*Pain in the ass*
PLZ	*Please*
PM	*Personal message*
PPL	*People*
Prblmz	*Problems*
Pwn	*Own (in the sense of controlling or dominating a situation, such as a game or argument. To be 'pwned' is to lose the game or argument, usually in a dramatic fashion.)*
Pwnage	*Ownage (The act of controlling or dominating a situation.)*
QT	*Cutie*
R	*Are*
Rily	*Really*
Rite	*Right*
ROFL	*Rolling on the floor laughing*
ROFLMAO	*Rolling on the floor laughing my ass off*
ROFLOL	*Roll on floor laughing out loud*
ROTFL	*Rolling on the floor laughing*
ROTFLMAO	*Rolling on the floor laughing my ass off*
ROTFLOL	*Rolling on the floor laughing out loud*
RTM	*Read the manual*
RU?	*Are you?*
Shez	*She's*
Shud	*Should*

SK8	*Skate*
Skul	*School*
Slf	*Self*
SME1	*Someone. Also, sm1*
Smtymz	*Sometimes*
Soz	*Sorry*
SSDD	*Same shit, different day*
STATS	*Your sex and age*
SUM1	*Someone*
Sumthing	*Something. Also, smthn*
SWTDRMZ	*Sweet dreams*
THANQ	*Thank you*
Therz	*There's*
Thn	*Then*
Thngz	*Things*
Thnk	*Think*
Thot	*Thought*
TIC	*Tongue in cheek*
Tok	*Talk, or took*
Toking	*Talking*
TTYL	*Talk to you later*
TY	*Thank you*
Tym	*Time*
TYVM	*Thank you very much*
U	*You*
U2	*You too. (Except when referring to the band of that name, in which case, U2.)*
U4E	*Yours forever*
Undrstnd	*Understand*
UR	*You are, or your*

Vry	*Very*
W8	*Wait*
WAN2	*Want to*
Wat	*What*
Wateva	*Whatever*
Waz	*Was*
WB	*Welcome back*
W/E	*Whatever*
Wen	*When. Also, Wn*
Weneva	*Whenever*
Wi	*We*
Wit	*With. Also, wif*
Wknd	*Weekend*
Wnt	*Want, or won't*
Wrk	*Work*
WTG	*Way to go!*
Wud	*Would*
WUD	*What you doing?*
Wyl	*While*
Wznt	*Wasn't*
XLNT	*Excellent*
Y	*Why*
ZZZ	*Sleeping, bored, tired*

Cn u rd Mxlish?

C f u cn understand thz msgs 4rm ppl using Mxit. Itz ez!
If you get stuck, use the Mxtionary.

A very cleaver poet said lv is nt lv bcz it dnt knw hw to lv, lv is nt lv bcz it dnt knw who 2 luv.

Wt a dae i had 2dy, met sm1 who i really like, she means a lot 2me, shes smart, sexy nd georgeous except shes a control freak. She jst askd me 2 marry hr, i mean days ago she tld me she wsnt ready 4 commitment nd nw dis. lyf sucks smtymz.

My daez are alwayz kwl, i mostly spend ma daez at da park wit ma friendz bt i do ma hme dutiez, lyk wash da dishez, clean cook or if itz skul daez i do ma hmeworkz and forget abt da park yea i love shoppin, bakin and djyein.

i wake at abt 8am sum mrning brash ma teeth grab a snack den hit da strts wit ma sk8board 4 a quike session til 10am afta i go bk hme do ma bed wash up eat brakfst den i hang out 4 a coupl of hourz watchin movies eatin up al da tym funy though i dnt get fat or enythng, wen my tym 2 go out is near i listen 2 ma sk8 inspirational musik watch sum sk8 videoz den get on da social netwrk.

1nc upn a tym dar wz ds guy i met on ma wy hme, he tld mi dt he lvs mi nd thn wi exchng numbrz thr afta wi clld each ada thn fingz developd4 1anada (wi d8td), he usd2 spoil mi buyn mi gfts, tykng mi2 fncy resturents etc. thn thr wz ds dai he wntd mi2 slp ova thn i refzd dwng dt thn he tld mi dt he nidz a brk coz m nt c. rus bwt ds, nd dt wz da end he nevr clld mi agn.